Crosswalk Coach for the Common Core State Standards

Mathematics

Grade 3

Crosswalk Coach for the Common Core State Standards, Mathematics, Grade 3
298NA
ISBN-13: 978-0-7836-7847-4

Contributing Writers: Cindy Frey and Andie Liao
Cover Image: © Ron Hilton/Dreamstime.com

Triumph Learning® 136 Madison Avenue, 7th Floor, New York, NY 10016

Printed in the United States of America.

20 19 18 17 16 15 14 13 12 11

Frequently Asked Questions about the Common Core State Standards

What are the Common Core State Standards?

The Common Core State Standards for mathematics and English language arts, grades K–12, are a set of shared goals and expectations for the knowledge and skills that will help students succeed. They allow students to understand what is expected of them and to become progressively more proficient in understanding and using mathematics and English language arts. Teachers will be better equipped to know exactly what they must do to help students learn and to establish individualized benchmarks for them.

Will the Common Core State Standards tell teachers how and what to teach?

No. Because the best understanding of what works in the classroom comes from teachers, these standards will establish *what* students need to learn, but they will not dictate *how* teachers should teach. Instead, schools and teachers will decide how best to help students reach the standards.

What will the Common Core State Standards mean for students?

The standards will provide a clear, consistent understanding of what is expected of student learning across the country. Common standards will not prevent different levels of achievement among students, but they will ensure more consistent exposure to materials and learning experiences through curriculum, instruction, teacher preparation, and other supports for student learning. These standards will help give students the knowledge and skills they need to succeed in college and careers.

Do the Common Core State Standards focus on skills and content knowledge?

Yes. The Common Core Standards recognize that both content and skills are important. They require rigorous content and application of knowledge through higher-order thinking skills. The English language arts standards require certain critical content for all students, including classic myths and stories from around the world, America's founding documents, foundational American literature, and Shakespeare. The remaining crucial decisions about content are left to state and local determination. In addition to content coverage, the Common Core State Standards require that students systematically acquire knowledge of literature and other disciplines through reading, writing, speaking, and listening.

In mathematics, the Common Core State Standards lay a solid foundation in whole numbers, addition, subtraction, multiplication, division, fractions, and decimals. Together, these elements support a student's ability to learn and apply more demanding math concepts and procedures.

The Common Core State Standards require that students develop a depth of understanding and ability to apply English language arts and mathematics to novel situations, as college students and employees regularly do.

Will common assessments be developed?

It will be up to the states: some states plan to come together voluntarily to develop a common assessment system. A state-led consortium on assessment would be grounded in the following principles: allowing for comparison across students, schools, districts, states and nations; creating economies of scale; providing information and supporting more effective teaching and learning; and preparing students for college and careers.

Table of Contents

Common Core State Standards Correlation Chart

Common Core State Standard	Grade 3	*Coach* Lesson(s)
Domain: Operations and Algebraic Thinking		
Represent and solve problems involving multiplication and division.		
3.OA.1	Interpret products of whole numbers, e.g., interpret 5×7 as the total number of objects in 5 groups of 7 objects each. *For example, describe a context in which a total number of objects can be expressed as* 5×7.	9
3.OA.2	Interpret whole-number quotients of whole numbers, e.g., interpret $56 \div 8$ as the number of objects in each share when 56 objects are partitioned equally into 8 shares, or as a number of shares when 56 objects are partitioned into equal shares of 8 objects each. *For example, describe a context in which a number of shares or a number of groups can be expressed as* $56 \div 8$.	16
3.OA.3	Use multiplication and division within 100 to solve word problems in situations involving equal groups, arrays, and measurement quantities, e.g., by using drawings and equations with a symbol for the unknown number to represent the problem.	9, 10, 12, 16–18
3.OA.4	Determine the unknown whole number in a multiplication or division equation relating three whole numbers. *For example, determine the unknown number that makes the equation true in each of the equations* $8 \times ? = 48$, $5 = \square \div 3$, $6 \times 6 = ?$.	9, 10, 16, 17
Understand properties of multiplication and the relationship between multiplication and division.		
3.OA.5	Apply properties of operations as strategies to multiply and divide. *Examples: If* $6 \times 4 = 24$ *is known, then* $4 \times 6 = 24$ *is also known. (Commutative property of multiplication.)* $3 \times 5 \times 2$ *can be found by* $3 \times 5 = 15$, *then* $15 \times 2 = 30$, *or by* $5 \times 2 = 10$, *then* $3 \times 10 = 30$. *(Associative property of multiplication.) Knowing that* $8 \times 5 = 40$ *and* $8 \times 2 = 16$, *one can find* 8×7 *as* $8 \times (5 + 2) = (8 \times 5) + (8 \times 2) = 40 + 16 = 56$. *(Distributive property.)*	13, 15
3.OA.6	Understand division as an unknown-factor problem. For example, find $32 \div 8$ by finding the number that makes 32 when multiplied by 8.	17
Multiply and divide within 100.		
3.OA.7	Fluently multiply and divide within 100, using strategies such as the relationship between multiplication and division (e.g., knowing that $8 \times 5 = 40$, one knows $40 \div 5 = 8$) or properties of operations. By the end of Grade 3, know from memory all products of two one-digit numbers.	10, 16, 17
Solve problems involving the four operations, and identify and explain patterns in arithmetic.		
3.OA.8	Solve two-step word problems using the four operations. Represent these problems using equations with a letter standing for the unknown quantity. Assess the reasonableness of answers using mental computation and estimation strategies including rounding.	5, 6, 8, 12, 18

Common Core State Standard	Grade 3	*Coach* Lesson(s)
Domain: Operations and Algebraic Thinking *(continued)*		
Solve problems involving the four operations, and identify and explain patterns in arithmetic. *(continued)*		
3.OA.9	Identify arithmetic patterns (including patterns in the addition table or multiplication table), and explain them using properties of operations. *For example, observe that 4 times a number is always even, and explain why 4 times a number can be decomposed into two equal addends.*	4, 11
Domain: Number and Operations in Base Ten		
Use place value understanding and properties of operations to perform multi-digit arithmetic.		
3.NBT.1	Use place value understanding to round whole numbers to the nearest 10 or 100.	1, 2, 7
3.NBT.2	Fluently add and subtract within 1000 using strategies and algorithms based on place value, properties of operations, and/or the relationship between addition and subtraction.	3, 5, 6
3.NBT.3	Multiply one-digit whole numbers by multiples of 10 in the range 10–90 (e.g., 9×80, 5×60) using strategies based on place value and properties of operations.	14
Domain: Number and Operations—Fractions		
Develop understanding of fractions as numbers.		
3.NF.1	Understand a fraction $\frac{1}{b}$ as the quantity formed by 1 part when a whole is partitioned into *b* equal parts; understand a fraction $\frac{a}{b}$ as the quantity formed by *a* parts of size $\frac{1}{b}$.	19
3.NF.2	Understand a fraction as a number on the number line; represent fractions on a number line diagram.	
3.NF.2.a	Represent a fraction $\frac{1}{b}$ on a number line diagram by defining the interval from 0 to 1 as the whole and partitioning it into *b* equal parts. Recognize that each part has size $\frac{1}{b}$ and that the endpoint of the part based at 0 locates the number $\frac{1}{b}$ on the number line.	19
3.NF.2.b	Represent a fraction $\frac{a}{b}$ on a number line diagram by marking off *a* lengths $\frac{1}{b}$ from 0. Recognize that the resulting interval has size $\frac{a}{b}$ and that its endpoint locates the number $\frac{a}{b}$ on the number line.	19
3.NF.3	Explain equivalence of fractions in special cases, and compare fractions by reasoning about their size.	
3.NF.3.a	Understand two fractions as equivalent (equal) if they are the same size, or the same point on a number line.	21
3.NF.3.b	Recognize and generate simple equivalent fractions, e.g., $\frac{1}{2} = \frac{2}{4}$, $\frac{4}{6} = \frac{2}{3}$. Explain why the fractions are equivalent, e.g., by using a visual fraction model.	21

Common Core State Standard	Grade 3	*Coach* Lesson(s)
Domain: Number and Operations—Fractions *(continued)*		
Develop understanding of fractions as numbers. *(continued)*		
3.NF.3.c	Express whole numbers as fractions, and recognize fractions that are equivalent to whole numbers. *Examples: Express 3 in the form* $3 = \frac{3}{1}$*; recognize that* $\frac{6}{1} = 6$*; locate* $\frac{4}{4}$ *and 1 at the same point of a number line diagram.*	20
3.NF.3.d	Compare two fractions with the same numerator or the same denominator by reasoning about their size. Recognize that comparisons are valid only when the two fractions refer to the same whole. Record the results of comparisons with the symbols $>$, $=$, or $<$, and justify the conclusions, e.g., by using a visual fraction model.	22
Domain: Measurement and Data		
Solve problems involving measurement and estimation of intervals of time, liquid volumes, and masses of objects.		
3.MD.1	Tell and write time to the nearest minute and measure time intervals in minutes. Solve word problems involving addition and subtraction of time intervals in minutes, e.g., by representing the problem on a number line diagram.	23
3.MD.2	Measure and estimate liquid volumes and masses of objects using standard units of grams (g), kilograms (kg), and liters (l). Add, subtract, multiply, or divide to solve one-step word problems involving masses or volumes that are given in the same units, e.g., by using drawings (such as a beaker with a measurement scale) to represent the problem.	24, 25
Represent and interpret data.		
3.MD.3	Draw a scaled picture graph and a scaled bar graph to represent a data set with several categories. Solve one- and two-step "how many more" and "how many less" problems using information presented in scaled bar graphs. *For example, draw a bar graph in which each square in the bar graph might represent 5 pets.*	30, 31
3.MD.4	Generate measurement data by measuring lengths using rulers marked with halves and fourths of an inch. Show the data by making a line plot, where the horizontal scale is marked off in appropriate units—whole numbers, halves, or quarters.	32, 33
Geometric measurement: understand concepts of area and relate area to multiplication and to addition.		
3.MD.5	Recognize area as an attribute of plane figures and understand concepts of area measurement.	
3.MD.5.a	A square with side length 1 unit, called "a unit square," is said to have "one square unit" of area, and can be used to measure area.	27
3.MD.5.b	A plane figure that can be covered without gaps or overlaps by *n* unit squares is said to have an area of *n* square units.	27

Common Core State Standard	Grade 3	*Coach* Lesson(s)
Domain: Measurement and Data *(continued)*		
Geometric measurement: understand concepts of area and relate area to multiplication and to addition. *(continued)*		
3.MD.6	Measure areas by counting unit squares (square cm, square m, square in, square ft, and improvised units).	27
3.MD.7	Relate area to the operations of multiplication and addition.	
3.MD.7.a	Find the area of a rectangle with whole-number side lengths by tiling it, and show that the area is the same as would be found by multiplying the side lengths.	28
3.MD.7.b	Multiply side lengths to find areas of rectangles with whole-number side lengths in the context of solving real world and mathematical problems, and represent whole-number products as rectangular areas in mathematical reasoning.	28
3.MD.7.c	Use tiling to show in a concrete case that the area of a rectangle with whole-number side lengths a and $b + c$ is the sum of $a \times b$ and $a \times c$. Use area models to represent the distributive property in mathematical reasoning.	28
3.MD.7.d	Recognize area as additive. Find areas of rectilinear figures by decomposing them into non-overlapping rectangles and adding the areas of the non-overlapping parts, applying this technique to solve real world problems.	28
Geometric measurement: recognize perimeter as an attribute of plane figures and distinguish between linear and area measures.		
3.MD.8	Solve real world and mathematical problems involving perimeters of polygons, including finding the perimeter given the side lengths, finding an unknown side length, and exhibiting rectangles with the same perimeter and different areas or with the same area and different perimeters.	26, 29
Domain: Geometry		
Reason with shapes and their attributes.		
3.G.1	Understand that shapes in different categories (e.g., rhombuses, rectangles, and others) may share attributes (e.g., having four sides), and that the shared attributes can define a larger category (e.g., quadrilaterals). Recognize rhombuses, rectangles, and squares as examples of quadrilaterals, and draw examples of quadrilaterals that do not belong to any of these subcategories.	34, 35
3.G.2	Partition shapes into parts with equal areas. Express the area of each part as a unit fraction of the whole. *For example, partition a shape into 4 parts with equal area, and describe the area of each part as* $\frac{1}{4}$ *of the area of the shape.*	36

Domain 1

Number and Operations in Base Ten

Domain 1: Diagnostic Assessment for Lessons 1–8

Domain 1: Cumulative Assessment for Lessons 1–8

Domain 1: Diagnostic Assessment for Lessons 1–8

1. Rich collected 3,567 cans to recycle. Regina collected 100 more cans than Rich. How many cans did Regina collect?

   ○ **A.** 4,567
   ○ **B.** 3,667
   ○ **C.** 3,557
   ○ **D.** 3,467

2. There were 4,293 people at the play on Saturday night. There were 3,872 people at the play on Sunday night. How many more people attended the play on Saturday than on Sunday?

   ○ **A.** 321
   ○ **B.** 421
   ○ **C.** 621
   ○ **D.** 1,421

3. Which list orders the numbers from least to greatest?

   ○ **A.** 2,357; 2,482; 2,498
   ○ **B.** 2,482; 2,357; 2,498
   ○ **C.** 2,498; 2,482; 2,357
   ○ **D.** 2,357; 2,498; 2,482

4. Mr. Turner's third-grade class baked 800 cookies for a bake sale. Mrs. Walter's third-grade class baked 300 more cookies than Mr. Turner's class. How many cookies did Mrs. Walter's class bake?

   ○ **A.** 500
   ○ **B.** 800
   ○ **C.** 1,000
   ○ **D.** 1,100

5. What is 3,871 − 1,000?

   ○ **A.** 4,871
   ○ **B.** 3,771
   ○ **C.** 2,871
   ○ **D.** 2,771

6. Which shows 428 rounded to the nearest ten?

   ○ **A.** 330
   ○ **B.** 400
   ○ **C.** 430
   ○ **D.** 530

7. Which number makes this sentence true?

$23 + 12 = \square + 23$

- ○ **A.** 11
- ○ **B.** 12
- ○ **C.** 23
- ○ **D.** 35

8. Ebony said that a good estimate for 412 + 381 is 800. Amber said it is 790. Who is correct?

- ○ **A.** Ebony is correct.
- ○ **B.** Amber is correct.
- ○ **C.** Neither is correct.
- ○ **D.** Both are correct.

9. What is the missing number in this pattern?

23 25 27 29 _?_ 33

____

10. Alexander and Ryan both collect baseball cards. Alexander has 387 cards in his collection. Ryan has 254 cards in his collection.

A. How many baseball cards do Alexander and Ryan have in all?

____

B. Tommy has 10 more cards than Alexander. How many cards does Tommy have?

____

Common Core State Standard:
3.NBT.1

Read and Write Whole Numbers

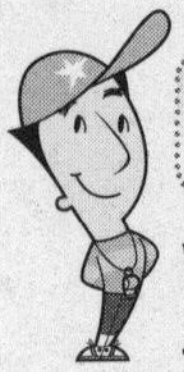

Getting the Idea

Whole numbers are the numbers 0, 1, 2, 3, 4, 5, and so on. **Digits** are used to write numbers.

The number 61,243 has five digits. Each digit's value is based on its position in the number. This is called its **place value**. A **place-value chart** can be used to show the value of each digit in a number.

Ten Thousands	Thousands		Hundreds	Tens	Ones
6	1	,	2	4	3

So 61,243 has 6 ten thousands, 1 thousand, 2 hundreds, 4 tens, and 3 ones.

The **place-value system** is based on 10s.

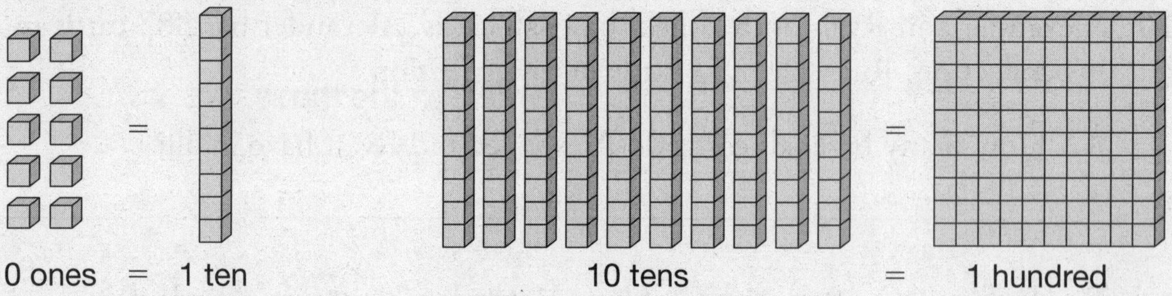

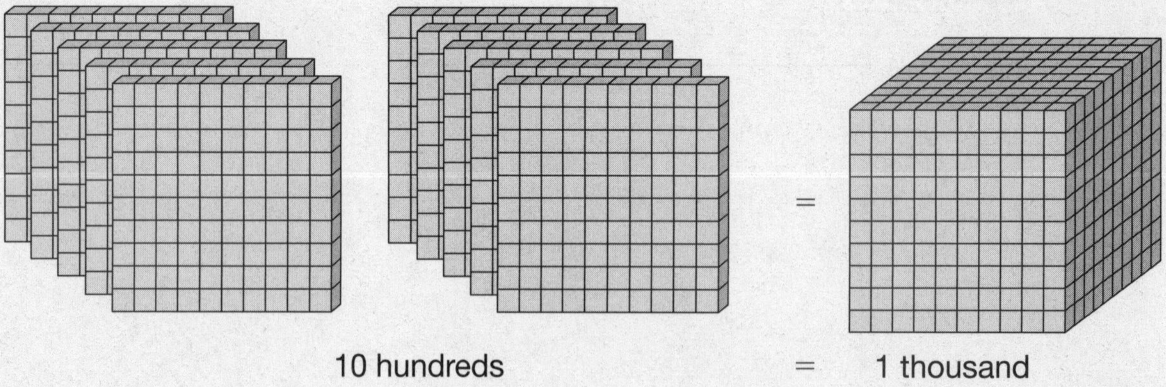

Example 1

What number do the models show?

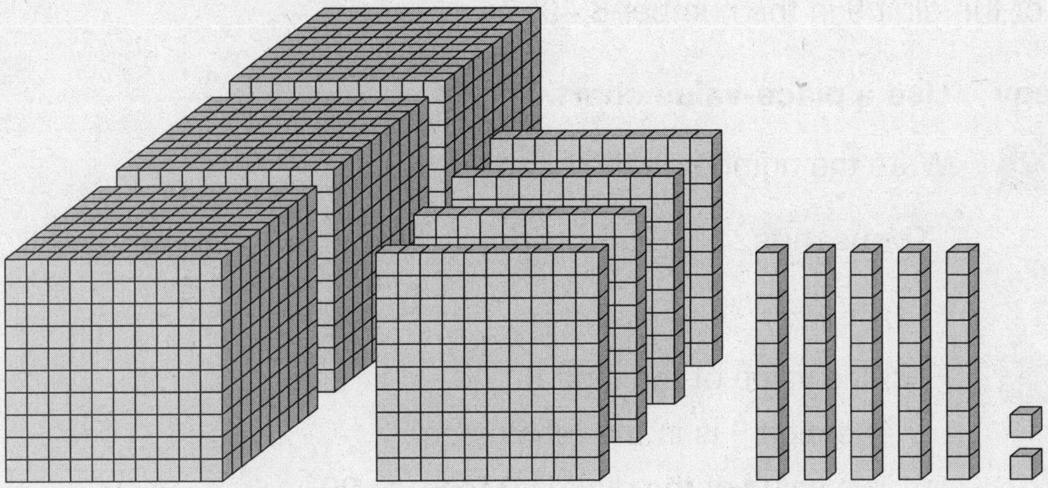

Strategy **Skip-count each group of blocks. Write the value of each group of blocks from greatest to least place-value position.**

Step 1 Skip-count the thousands.

1,000; 2,000; 3,000

There are 3 thousands. Write 3 in the thousands place and a comma to separate the thousands and hundreds places.

3,

Step 2 Skip-count the hundreds.

100, 200, 300, 400

There are 4 hundreds. Write a 4 in the hundreds place.

3,4

Step 3 Skip-count the tens.

10, 20, 30, 40, 50

There are 5 tens. Write a 5 in the tens place.

3,45

Step 4 Count the ones.

There are 2 ones. Write a 2 in the ones place.

3,452

Solution **The models show the number 3,452.**

Example 2

A singer's new song was downloaded 8,495 times in one day. What is the value of the digit 9 in the number 8,495?

Strategy **Use a place-value chart.**

Step 1 Write the number in the chart.

Thousands		Hundreds	Tens	Ones
8	,	4	9	5

Step 2 Find the value of the digit 9.

The digit 9 is in the tens place.

The value of the digit is 9 tens, or 90.

Solution **The value of the digit 9 in 8,495 is 9 tens, or 90.**

Numbers can be written in different forms.

Base-ten numerals: 28,495

Number name: twenty-eight thousand, four hundred ninety-five

Expanded form: 20,000 + 8,000 + 400 + 90 + 5

Example 3

What is the number name for 81,173?

Strategy **Use place value.**

Step 1 Read the value of the digits to the left of the comma.

81 thousands = 81,000

Write 81,000 in words, then insert a comma.

eighty-one thousand,

Step 2 Read the value of the digits to the right of the comma.

1 hundred, 7 tens, 3 ones = 173

Write 173 in words.

one hundred seventy-three

Step 3 Write the number name.

eighty-one thousand, one hundred seventy-three

Solution **The number name for 81,173 is eighty-one thousand, one hundred seventy-three.**

Example 4

A library has 19,827 books. How do you write 19,827 in expanded form?

Strategy **Find the value of each digit. Then list the values together.**

Step 1 Find the value of each digit.

1 ten thousand = 10,000

9 thousands = 9,000

8 hundreds = 800

2 tens = 20

7 ones = 7

Step 2 Write the expanded form. Use a + between each value.

10,000 + 9,000 + 800 + 20 + 7

Solution **Written in expanded form, the number 19,827 is 10,000 + 9,000 + 800 + 20 + 7.**

5,235.

↓

5,000 + 200 + 30 + 5

Coached Example

What number do the models show?

3,000 + 600 + 20 + 4

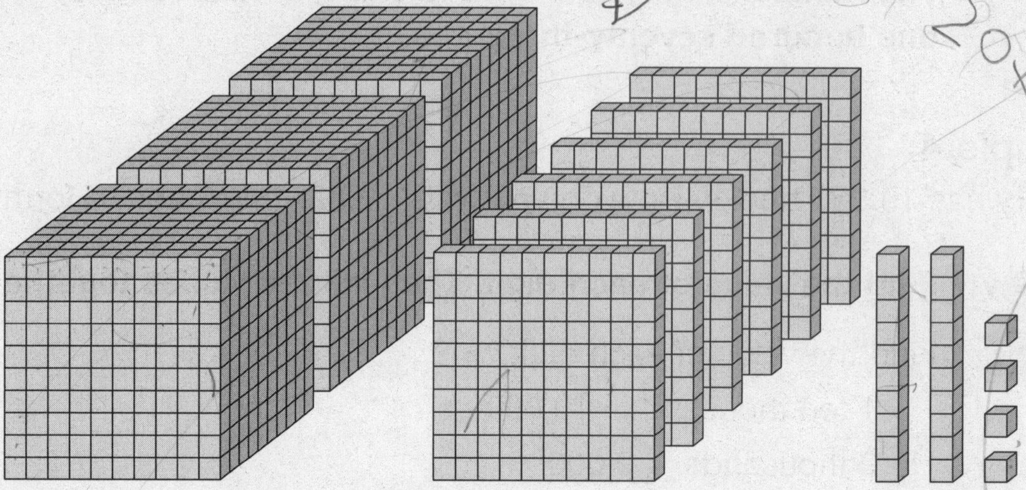

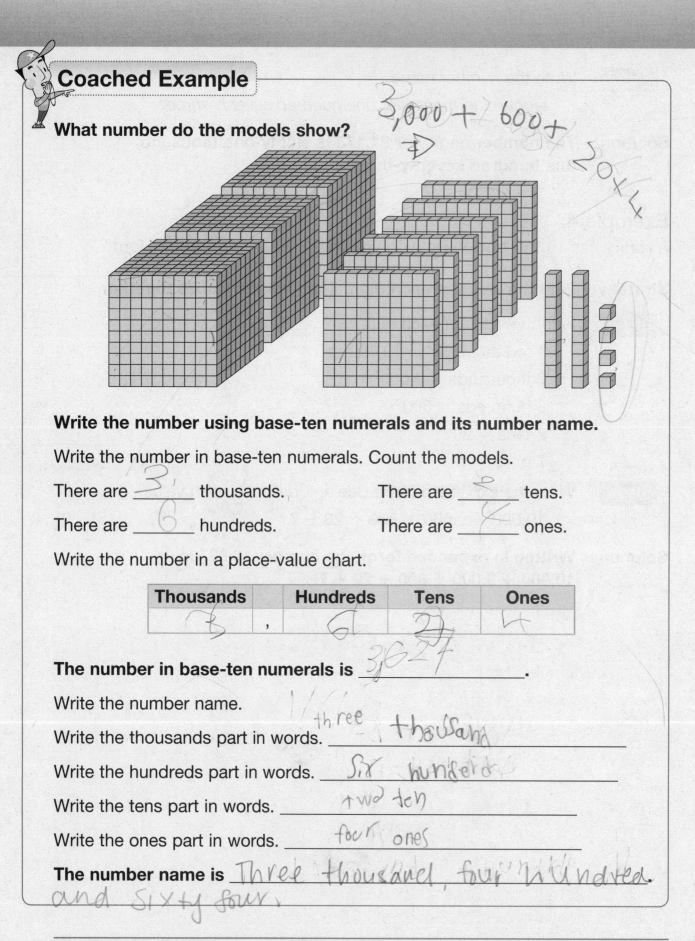

Write the number using base-ten numerals and its number name.

Write the number in base-ten numerals. Count the models.

There are __3__ thousands. There are __2__ tens.

There are __6__ hundreds. There are __4__ ones.

Write the number in a place-value chart.

Thousands		Hundreds	Tens	Ones
3	,	6	2	4

The number in base-ten numerals is __3,624__**.**

Write the number name.

Write the thousands part in words. __three thousand__

Write the hundreds part in words. __six hundred__

Write the tens part in words. __two ten__

Write the ones part in words. __four ones__

The number name is __Three thousand, four hundred and sixty four.__**.**

Lesson Practice

Choose the correct answer.

1. Which shows 1 thousand, 5 hundreds, 9 tens, 0 ones?

- ○ **A.** 159
- ○ **B.** 951
- ○ **C.** 1,509
- ○ **D.** 1,590

2. Which is the number name for 2,215?

- ○ **A.** two thousand, one hundred twenty-five
- ○ **B.** two thousand, two hundred fifteen
- ○ **C.** five thousand, one hundred twenty-two
- ○ **D.** twenty-two hundred, fifty-five

3. Which number has a 6 in the tens place?

- ○ **A.** 1,768
- ○ **B.** 2,316
- ○ **C.** 4,625
- ○ **D.** 6,184

4. Which shows the number forty-five thousand, eight-hundred nineteen in base-ten numerals?

- ○ **A.** 45,819
- ○ **B.** 45,189
- ○ **C.** 48,591
- ○ **D.** 91,854

5. Which is the expanded form of 62,803?

- ○ **A.** 6,000 + 200 + 80 + 3
- ○ **B.** 60,000 + 2,000 + 80 + 3
- ○ **C.** 60,000 + 2,000 + 800 + 3
- ○ **D.** 60,000 + 2,000 + 800 + 30

6. What number is shown by the models?

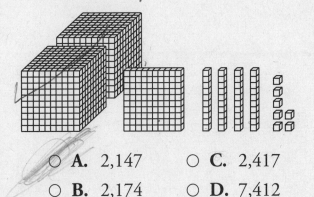

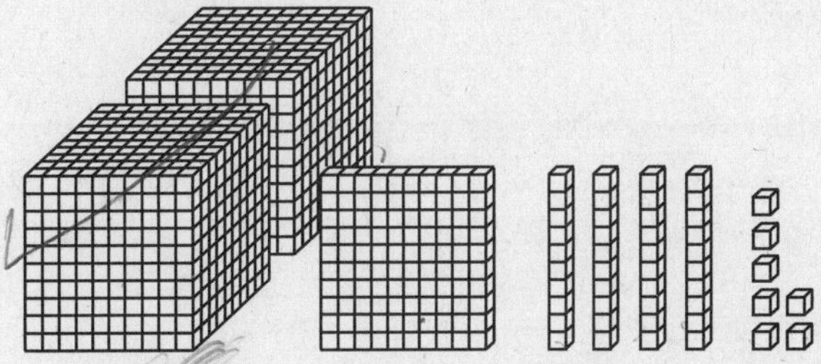

- ○ **A.** 2,147
- ○ **B.** 2,174
- ○ **C.** 2,417
- ○ **D.** 7,412

7. There are 38,072 seats in a stadium. What is the number name for 38,072?

- ○ **A.** thirty-eight thousand, twenty-seven
- ○ **B.** thirty-eight thousand, seven hundred twenty
- ○ **C.** thirty-eight thousand, seven hundred two
- ○ **D.** thirty-eight thousand, seventy-two

8. Frank modeled the number below.

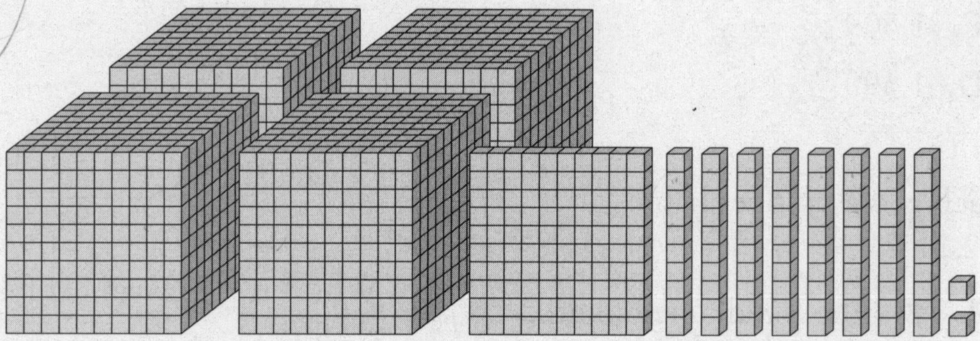

A. Write the number in base-ten numerals.

____

B. Write the number name.

____

C. Write the number in expanded form.

____

Common Core State Standard:
3.NBT.1

Compare and Order Whole Numbers

Getting the Idea

You can compare and order whole numbers by looking at their place values. To compare numbers, use the following symbols:

$>$ means **is greater than.**

$<$ means **is less than.**

$=$ means **is equal to.**

Example 1

Which symbol makes this statement true? Write $>$, $<$, or $=$.

5,358 ◯ 5,385

Strategy **Line up the numbers on the ones place. Then compare the digits from left to right.**

Step 1 Line up the digits on the ones place.

5,358
5,385

Step 2 Compare the greatest place: thousands.

5,358
5,385

Since 5 = 5, compare the next greatest place: hundreds.

Step 3 Compare the hundreds.

5,**3**58
5,**3**85

Since 3 = 3, compare the next greatest place: tens.

Step 4 Compare the tens.

5,358

5,385

5 < 8

So 5,358 is less than 5,385. Use the symbol <.

Solution **5,358 (<) 5,385**

Example 2

Compare. Use >, <, or =.

68,471 ◯ 68,459

Strategy **Use a place-value chart to compare the numbers.**

Step 1 Write the numbers in a place-value chart.

Ten Thousands	Thousands		Hundreds	Tens	Ones
6	8	,	4	7	1
6	8	,	4	5	9

Step 2 Compare the digits in the greatest place: ten thousands.

6 ten thousands = 6 ten thousands

Since the ten thousands are equal, compare the thousands.

Step 3 Compare the thousands.

8 thousands = 8 thousands

Since the thousands are equal, compare the hundreds.

Step 4 Compare the hundreds.

4 hundreds = 4 hundreds

Since the hundreds are equal, compare the tens.

Step 5 Compare the tens.

7 tens > 5 tens

The greater number is 68,471, so use the symbol >.

Solution **68,471 (>) 68,459**

Example 3

The table below shows the highest elevations in four U.S. states.

Highest Elevations

State	Elevation (in feet)
Georgia	4,784
Oklahoma	4,973
Vermont	4,393
West Virginia	4,863

Order these states from greatest to least elevation.

Strategy **Line up the numbers on the ones place. Compare the digits from left to right.**

Step 1 Line up the digits on the ones place.

4,784
4,973
4,393
4,863

Step 2 Compare the digits in the greatest place: thousands.

All the digits have a 4 in the thousands place.

Move on to the hundreds place.

Step 3 Compare the hundreds place of the numbers.

9 is greater than 8, 7, or 3. So 4,973 is the greatest number.

8 is greater than 7 or 3. So 4,863 is the second greatest number.

7 is greater than 3. So 4,784 is the next greatest number.

3 is less than 9, 8, or 7. So 4,393 is the least number.

Step 4 Order the numbers from greatest to least.

4,973; 4,863; 4,784; 4,393

Step 5 Replace the numbers in the list with their states.

Oklahoma, West Virginia, Georgia, Vermont

Solution **From greatest to least elevation, the order of the states is Oklahoma, West Virginia, Georgia, and Vermont.**

Coached Example

Order the following numbers from greatest to least.

7,736 **7,175** **7,742**

Write the numbers by lining up the digits on the ones place.

____

____

____

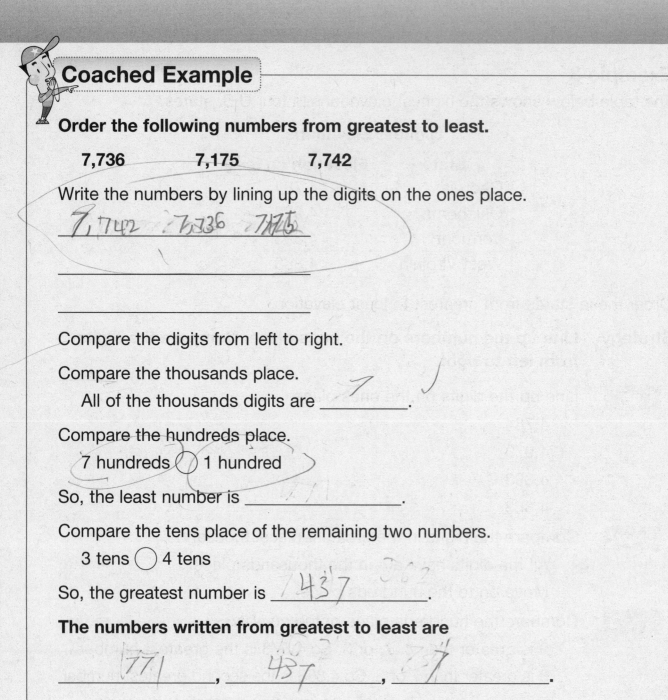

Compare the digits from left to right.

Compare the thousands place.

All of the thousands digits are ________.

Compare the hundreds place.

7 hundreds ◯ 1 hundred

So, the least number is ______________.

Compare the tens place of the remaining two numbers.

3 tens ◯ 4 tens

So, the greatest number is ______________.

The numbers written from greatest to least are

______________, ______________, ______________.

Lesson Practice

Choose the correct answer.

1. Which number has the least value?
   - A. 73,629
   - B. 73,926
   - C. 73,692
   - D. 73,296

2. Which of the following is true?
   - A. $6{,}507 = 6{,}570$
   - B. $7{,}125 > 7{,}152$
   - C. $7{,}439 < 7{,}451$
   - D. $9{,}381 < 9{,}318$

3. Which list orders the numbers from least to greatest?
   - A. 7,257 7,527 7,725
   - B. 7,725 7,257 7,527
   - C. 7,527 7,257 7,725
   - D. 7,527 7,725 7,257

4. Which number has the greatest value?
   - A. 82,157
   - B. 83,526
   - C. 83,265
   - D. 82,751

5. The table shows the number of people in four Kentucky cities.

City Populations

City	Number of People
Alexandria	8,286
Benton	4,197
Central City	5,893
Flatwoods	7,605

Which of the following cities has the second greatest number of people?
   - A. Alexandria
   - B. Benton
   - C. Central City
   - D. Flatwoods

6. Which digit makes this sentence true?

53,426 < 53,☐09

○ **A.** 2

○ **B.** 3

○ **C.** 4

○ **D.** 5

7. Which number is greater than 14,520 and less than 14,549?

○ **A.** 14,508

○ **B.** 14,630

○ **C.** 14,497

○ **D.** 14,532

8. Which of the following is **not** true?

○ **A.** 62,749 > 62,801

○ **B.** 31,597 > 31,499

○ **C.** 85,151 < 85,164

○ **D.** 96,156 < 96,463

9. Ms. Frankel wrote the following two numbers on the board.

43,892 43,928

She asked the class to compare the numbers.

A. Write the symbol to make this sentence true. Use >, <, or =.

43,892 ○ 43,928

B. Ms. Frankel wrote a third number on the board. Order Ms. Frankel's numbers from least to greatest.

43,892 43,928 43,392

____

Common Core State Standard:
3.NBT.2

Addition Properties

Getting the Idea

Using addition properties can make it easier for you to add numbers.

The **commutative property of addition** says that changing the order of the addends does not change the sum.

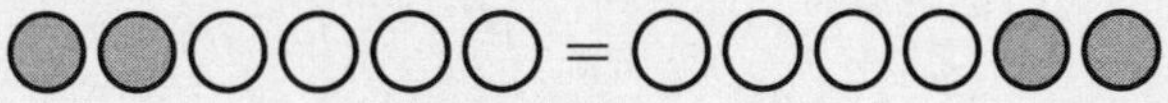

$$2 + 4 = 4 + 2$$
$$6 = 6$$

Example 1

Which number makes this sentence true?

$12 + 18 = 18 + \square$

Strategy **Use the commutative property of addition.**

The commutative property of addition says that changing the order of the addends does not change the sum.

$12 + 18 = 30$

$18 + 12 = 30$

Solution **The number 12 makes the sentence true.**

The **associative property of addition** says that changing the grouping of the addends does not change the sum.

For example, $(4 + 3) + 5 = 4 + (3 + 5)$.

$(4 + 3) + 5 = 7 + 5$ $\qquad$ $4 + (3 + 5) = 4 + 8$

$7 + 5 = 12$ $\qquad$ $4 + 8 = 12$

Example 2

Find the sum.

$3 + (17 + 6) = \square$

Strategy **Use the associative property of addition.**

Step 1 Change the grouping of the addends.

Use mental math.

Think: $3 + 17 = 20$

$3 + (17 + 6) = (3 + 17) + 6$

Step 2 Find the sum.

$(3 + 17) + 6$

$20 \quad + 6 = 26$

Solution $\mathbf{3 + (17 + 6) = 26}$

The **identity property of addition** says that the sum of any addend and 0 is that addend.

For example, $9 + 0 = 9$ and $0 + 17 = 17$.

Example 3

What number goes in the □ to make the sentence true?

0 + □ = 23

Strategy **Use the identity property of addition.**

Step 1 Look at the numbers in the number sentence 0 + □ = 23.

The addend is 0. The sum is 23.

Step 2 Find the other addend.

Use the identity property of addition.

When an addend is added to 0, the sum is the addend.

0 + 23 = 23

Solution **0 + 23 = 23**

Coached Example

If 4 + 9 = 13, what is the missing addend in the number sentence below?

□ + 4 = 13

What is the sum in the number sentence 4 + 9 = 13? ________

What is the sum in the number sentence □ + 4 = 13? ________

Are the sums the same? ________________

What are the two addends in the number sentence 4 + 9 = 13?

________ and ________

What property of addition says that adding the addends in a different order does not change the sum? ________________________

What is the missing addend in the number sentence □ + 4 = 13? ________

The missing addend is ________________.

Lesson Practice

Choose the correct answer.

1. Which number makes this sentence true?

$15 + 16 = \square + 15$

- **A.** 31
- **B.** 21
- **C.** 16
- **D.** 15

2. Which number makes this sentence true?

$4 + \square = 4$

- **A.** 0
- **B.** 1
- **C.** 4
- **D.** 5

3. Which number makes this sentence true?

$\square + 7 = 7$

- **A.** 0
- **B.** 1
- **C.** 7
- **D.** 14

4. Which is the missing number?

$(22 + 17) + 3 = 22 + (17 + \square)$

- **A.** 42
- **B.** 22
- **C.** 17
- **D.** 3

5. Which number makes the sentence true?

$\square + 5 = 5 + 4$

- **A.** 0
- **B.** 1
- **C.** 4
- **D.** 9

6. Which is the missing number?

$(3 + \square) + 5 = 3 + (4 + 5)$

- **A.** 3
- **B.** 4
- **C.** 5
- **D.** 12

7. Which number makes the sentence true?

$0 + 8 = \square$

- ○ **A.** 0
- ○ **B.** 1
- ○ **C.** 8
- ○ **D.** 9

8. Sasha wants to add the numbers below.

$9 + (1 + 7) = \square$

What property of addition could be used to find the sum?

- ○ **A.** commutative property
- ○ **B.** associative property
- ○ **C.** identity property
- ○ **D.** zero property

9. Jamie said that the following is a true number sentence.

$13 + (4 + 7) = 13 + (7 + 4)$

A. Which property of addition makes the sentence true?

____

B. Jamie then wrote this number sentence. Which property of addition makes this sentence true?

$13 + (7 + 4) = (13 + 7) + 4$

____

C. What is the sum? Show your work.

$(13 + 7) + 4$

____

Common Core State Standard:
3.OA.9

Patterns

Getting the Idea

A **number pattern** is a group of numbers that follows a rule. The rule describes how the numbers are related.

The numbers in this pattern increase.

Rule: Add 5.

20 25 30 35 40 45

The numbers in this pattern decrease.

Rule: Subtract 2.

60 58 56 54 52 50

Example 1

What is the next number in this pattern?

3 6 9 12 ___?___

Strategy **Decide if the numbers increase or decrease. Find the rule of the pattern.**

Step 1 Do the numbers increase or decrease?

The numbers increase.

Step 2 Find how many are between the first two numbers in the pattern.

Think: $3 + ? = 6$

$3 + \mathbf{3} = 6$

Try adding 3 to each number.

Step 3 Find the rule.

$3 + \mathbf{3} = 6$

$6 + \mathbf{3} = 9$

$9 + \mathbf{3} = 12$

Each number is 3 more than the number before it.

The rule is to add 3.

Step 4 Find the next number in the pattern.

Use the rule. Add 3 to 12.

12 + **3** = 15

Solution The next number in this pattern is 15.

You can find many patterns in an addition table.

For example, the sums in each row increase by 1 as you go from left to right.

+	1	2	3	4	5	6	7	8	9	10
1	**2**	3	4	5	6	7	8	9	10	11
2	3	**4**	5	6	7	8	9	10	11	12
3	4	5	**6**	7	8	9	10	11	12	13
4	5	6	7	**8**	9	10	11	12	13	14
5	6	7	8	9	**10**	11	12	13	14	15
6	7	8	9	10	11	**12**	13	14	15	16
7	8	9	10	11	12	13	**14**	15	16	17
8	9	10	11	12	13	14	15	**16**	17	18
9	10	11	12	13	14	15	16	17	**18**	19
10	11	12	13	14	15	16	17	18	19	**20**

There are even number patterns and odd number patterns.

An **even number** can be separated into two equal groups.
An even number has 0, 2, 4, 6, or 8 in the ones place.

An **odd number** has 1 left over after being separated into two equal groups.
An odd number has 1, 3, 5, 7, or 9 in the ones place.

When adding a number to itself, the sum is always an even number.
For example, 1 + 1 = 2, 2 + 2 = 4, 3 + 3 = 6, and so on.
The boldface numbers in the table show this pattern.

When adding an even number to an odd number, the sum is always an odd number. For example, 3 + 6 = 9 and 7 + 4 = 11.

Example 2

Julia saw these mailboxes along one side of Poplar Street.

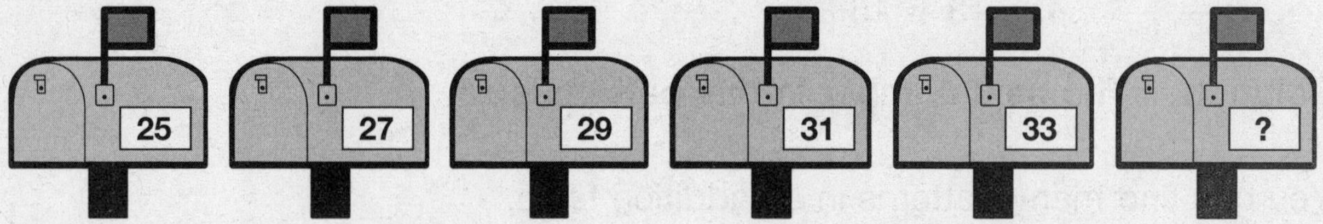

Which is most likely to be the number on the next mailbox?

Are the numbers on the mailboxes odd or even?

Strategy **Find the rule of the pattern.**

Step 1 Do the numbers increase or decrease?

The numbers increase.

Step 2 Find how many are between the first two numbers in the pattern.

Think: 25 + ? = 27

25 + **2** = 27

Try adding 2 to each number.

Step 3 Find the rule.

25 + **2** = 27

27 + **2** = 29

29 + **2** = 31

31 + **2** = 33

Each number is 2 more than the number before it.

The rule is to add 2.

Step 4 Use the rule to find the next number.

33 + **2** = 35

Step 5 Decide if the numbers are odd or even.

All of the numbers have 1, 3, 5, 7, or 9 in the ones place.

So, the numbers are odd numbers.

Solution **The number on the next mailbox is most likely 35.**
The numbers are all odd numbers.

Example 3

What is the missing number in this pattern?

22 18 14 10 ? 2

Strategy **Find the rule of the pattern.**

Step 1 Do the numbers increase or decrease?

The numbers decrease.

Step 2 Find how many are between the first two numbers in the pattern.

Think: $22 - ? = 18$

$22 - \mathbf{4} = 18$

Try subtracting 4 from each number.

Step 3 Find the rule.

$22 - \mathbf{4} = 18$

$18 - \mathbf{4} = 14$

$14 - \mathbf{4} = 10$

Each number is 4 less than the number before it.

The rule is to subtract 4.

Step 4 Use the rule to find the missing number.

Subtract 4 from 10.

$10 - \mathbf{4} = 6$

Step 5 Check to make sure the missing number is correct.

Subtract 4 from 6 to make sure the next number is 2.

$6 - \mathbf{4} = 2$

Solution **The missing number in this pattern is 6.**

Coached Example

What is the next number in this pattern?

34 31 28 25 22 ?

Are the numbers in the pattern odd or even?

Do the numbers in the pattern increase or decrease? ________

Find how many are between the first two numbers.

31 is ________ less than 34.

Try subtracting ________ from each number.

34 − ________ = ________

31 − ________ = ________

28 − ________ = ________

25 − ________ = ________

The rule is ______________________.

Use the rule to find the next number in the pattern.

22 − ________ = ________

The next number in the pattern is ________.

Decide if the numbers are odd or even.

Look at the ____________ digit in each number.

The even numbers in the pattern are ________, ________, and ________.

The odd numbers in the pattern are ________, ________, and ________.

The numbers in the pattern are both ____________ and ____________.

Lesson Practice

Choose the correct answer.

1. Which is the next number in this pattern?

   1 3 5 7 9 ?

   - A. 8
   - B. 10
   - C. 11
   - D. 12

2. Which is the next number in this pattern?

   4 8 12 16 20 ?

   - A. 18
   - B. 21
   - C. 22
   - D. 24

3. Marcus made a pattern using the rule add 5. Which could be Marcus's number pattern?

   - A. 5 10 14 17 19 20
   - B. 6 11 16 21 26 31
   - C. 7 11 15 19 23 27
   - D. 8 13 18 24 30 35

Use this number pattern for questions 4 and 5.

17 15 13 11 ? 7

4. What is the missing number in the pattern?

   - A. 12
   - B. 10
   - C. 9
   - D. 8

5. Which is true about the numbers in the pattern?

   - A. All of the numbers are odd numbers.
   - B. All of the numbers are even numbers.
   - C. Three of the numbers are odd numbers.
   - D. Only one number is an even number.

6. Which is the next number in the pattern?

   7 13 19 25 31 ?

   - A. 33
   - B. 35
   - C. 37
   - D. 41

Use this number pattern for questions 7 and 8.

37 34 31 28 ? 22

7. Which is the missing number in the pattern?

   ○ **A.** 27

   ○ **B.** 25

   ○ **C.** 24

   ○ **D.** 23

8. Which is true about the numbers in this pattern?

   ○ **A.** All of the numbers are even numbers.

   ○ **B.** All of the numbers are odd numbers.

   ○ **C.** Only one number is an even number.

   ○ **D.** Three of the numbers are odd numbers.

9. Maureen began a workout routine. Her workout was 15 minutes on the first day, 18 minutes on the second day, 21 minutes on the third day, 24 minutes on the fourth day, and 27 minutes on the fifth day.

   A. The pattern of the workout minutes is 15, 18, 21, 24, 27.

   If the pattern continues, how many minutes will Maureen's workout be on the sixth day? Explain your answer.

   ____

   ____

   ____

   B. Are the numbers of workout minutes odd or even? Explain your answer.

   ____

   ____

   ____

Common Core State Standards: 3.OA.8, 3.NBT.2

Add Whole Numbers

Getting the Idea

When you **add**, you combine quantities.

Here are the parts to an addition sentence.

43	+	25	=	68
addend		**addend**		**sum**

You can write an addition problem in a column. Line up the digits on the ones place. Then add from right to left. When the sum of a column is 10 or greater, **regroup** 10 of one unit to 1 of the next greatest unit.

Example 1

Keisha has 231 trading cards. Her brother has 467 trading cards.

How many trading cards do they have in all?

Strategy **Add to find how many cards in all.**

Step 1 Write an addition sentence for the problem.

Use the symbol □ for the sum.

231 + 467 = □

Step 2 Write the addition problem in a column. Line up the addends by place value.

$$\begin{array}{r} 231 \\ +\ 467 \\ \hline \end{array}$$

Step 3 Add the ones.

1 one + 7 ones = 8 ones

$$\begin{array}{r} 23\mathbf{1} \\ +\ 46\mathbf{7} \\ \hline \mathbf{8} \end{array}$$

Step 4 Add the tens.

3 tens + 6 tens = 9 tens

$$\begin{array}{r} 2\mathbf{3}1 \\ +\,4\mathbf{6}7 \\ \hline \mathbf{9}8 \end{array}$$

Step 5 Add the hundreds.

2 hundreds + 4 hundreds = 6 hundreds

$$\begin{array}{r} \mathbf{2}31 \\ +\,\mathbf{4}67 \\ \hline \mathbf{6}98 \end{array}$$

Solution **Keisha and her brother have 698 trading cards in all.**

Example 2

What is the sum of 524 + 197?

Strategy **Rewrite the addition problem in a column. Then add from right to left.**

Step 1 Add the ones.

4 ones + 7 ones = 11 ones

Regroup 11 ones as 1 ten and 1 one.

$$\begin{array}{r} \mathbf{1} \\ 52\mathbf{4} \\ +\,19\mathbf{7} \\ \hline \mathbf{1} \end{array}$$

Step 2 Add the tens. Remember to include the 1 regrouped ten.

1 ten + 2 tens + 9 tens = 12 tens

Regroup 12 tens as 1 hundred and 2 tens.

$$\begin{array}{r} \mathbf{11} \\ 5\mathbf{2}4 \\ +\,1\mathbf{9}7 \\ \hline \mathbf{2}1 \end{array}$$

Step 3 Add the hundreds. Remember to include the 1 regrouped hundred.

1 hundred + 5 hundreds + 1 hundred = 7 hundreds

$$\begin{array}{r} {\scriptstyle 11} \\ \mathbf{5}24 \\ +\ \mathbf{1}97 \\ \hline \mathbf{7}21 \end{array}$$

Solution **524 + 197 = 721**

You can use mental math to add numbers that end in 0. Use place value.

For example:

620 + 10 =
620 + 1 ten = 630

620 + 300 =
620 + 3 hundreds = 920

Example 3

What is 400 + 200?

Strategy **Use mental math.**

Step 1 Think: How many hundreds in 200?

There are 2 hundreds in 200.

Step 2 Think: Which digit will change in 400 when I add 2 hundreds?

4 is the digit in the hundreds place. It will increase by 2.

Step 3 Mentally add.

400 + 200 (or 2 hundreds) = 600

Solution **400 + 200 = 600**

Example 4

Lilly collected 283 plastic bottles for her school's recycling drive. Darren collected 100 more than Lilly. How many plastic bottles did Darren collect?

283 + 100 = □ 383

Strategy **Use mental math.**

Step 1 Think: How many hundreds in 100?

There is 1 hundred in 100.

Step 2 Think: Which digit will change in 283 when I add 1 hundred?

2 is the digit in the hundreds place. It will increase by 1.

Step 3 Add mentally.

283 + 100 (or 1 hundred) = 383

Solution **Darren collected 383 plastic bottles.**

Coached Example

Rhea has 274 pennies in her jar. Tyler has 400 more pennies in his jar. How many pennies does Tyler have in his jar?

274 + 400 = □

Use mental math.

How many hundreds in 400? 4 hunderd

Which digit will change in 274 when you add 4 hundreds?

The digit 1 is in the hundreds place. It will increase by 1.

Mentally add.

274 + 400 = 674

Tyler has 674 **pennies in his jar.**

Lesson Practice

Choose the correct answer.

1. What is 156 + 415?

- ○ **A.** 571
- ○ **B.** 561
- ○ **C.** 541
- ○ **D.** 341

2. Samantha is reading a book that is 631 pages long. Frank is reading a book that is 10 pages longer. How many pages are in Frank's book?

- ○ **A.** 621
- ○ **B.** 632
- ○ **C.** 641
- ○ **D.** 731

3. What is the sum?

$$\begin{array}{r} 215 \\ +\ 498 \\ \hline \end{array}$$

- ○ **A.** 703
- ○ **B.** 713
- ○ **C.** 714
- ○ **D.** 723

4. A town has two schools. There are 473 students at one school. The other school has 354 students. How many students are there at both schools?

- ○ **A.** 727
- ○ **B.** 827
- ○ **C.** 837
- ○ **D.** 927

5. A band played two concerts. The first concert had 375 people. The second concert had 200 more people than the first concert. How many people attended the second concert?

- ○ **A.** 575
- ○ **B.** 475
- ○ **C.** 395
- ○ **D.** 385

6. Find the sum.

436 + 115 = □

- ○ **A.** 541
- ○ **B.** 551
- ○ **C.** 578
- ○ **D.** 587

7. Roger played a computer game twice. He scored 428 points in his first game. In the second game, he scored 559 points. How many points did Roger score in all?

- ○ **A.** 887
- ○ **B.** 967
- ○ **C.** 977
- ○ **D.** 987

8. What is 300 + 500?

- ○ **A.** 200
- ○ **B.** 700
- ○ **C.** 800
- ○ **D.** 900

9. Ethel and Patsy have sticker collections. Ethel has 645 stickers in her collection. Patsy has 289 stickers in her collection.

A. How many stickers do Ethel and Patsy have in all?

____

B. Betty has 10 more stickers than Ethel. How many stickers does Betty have?

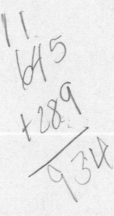

____

Common Core State Standards: 3.OA.8, 3.NBT.2

Subtract Whole Numbers

Getting the Idea

When you **subtract**, you take away from a quantity.

Here are the parts of a subtraction sentence.

479	−	236	=	243
minuend		**subtrahend**		**difference**

You can write a subtraction problem in a column. Line up the digits on the ones place. Then subtract from right to left. Sometimes you may need to **regroup**.

Example 1

Jeff invited 62 guests to his party. 28 guests are adults. The rest are children. How many children did Jeff invite to his party?

Strategy **Subtract to find how many guests are children.**

Step 1 Write a subtraction sentence for the problem.

Use the symbol ▭ for the difference.

$62 - 28 = \square$

Step 2 Write the subtraction problem in a column. Line up the digits on the ones place.

Since 8 is greater than 2, regroup 1 ten as 10 ones.

$$\begin{array}{r} {}^{5}\,{}^{12} \\ \not{6}\,\not{2} \\ -\,2\,8 \\ \hline \end{array}$$

Step 3 Subtract the ones.

12 ones − 8 ones = 4 ones

$$\begin{array}{r} {}^{5}\,{}^{12} \\ \not{6}\,\not{2} \\ -\,2\,8 \\ \hline 4 \end{array}$$

Step 4 Subtract the tens.

5 tens − 2 tens = 3 tens

$$\begin{array}{r} \mathbf{5}\;12 \\ \not{6}\;\not{2} \\ -\;\mathbf{2}\;8 \\ \hline \mathbf{3}\;4 \end{array}$$

Solution **Jeff invited 34 children to his party.**

You can use addition to check the difference.

$$\begin{array}{r} 62 \\ -\;28 \\ \hline 34 \end{array} \quad \searrow \quad \begin{array}{r} 34 \\ +\;28 \\ \hline 62 \end{array}$$

Example 2

Aaron is reading a trilogy. He read 346 pages of the second book. He has 487 more pages to read to finish the book. The first book has 637 pages. How many more pages does the second book have than the first book?

Strategy **First, add to find the number of pages in the second book. Then subtract to find how many more pages are in the second book.**

Step 1 Write an addition sentence to find the number of pages in the second book.

346 + 487 = □

Add the ones, then the tens, then the hundreds.
Regroup as needed.

$$\begin{array}{r} {\scriptstyle 1\,1}\;\; \\ 346 \\ +\;487 \\ \hline 833 \end{array}$$

Aaron's second book has 833 pages.

Step 2 Write a subtraction sentence to find how many more pages are in the second book.

833 − 637 = □

Subtract the ones, then the tens, then the hundreds.
Regroup as needed.

$$\begin{array}{r} {\scriptstyle 12} \\ {\scriptstyle 7\ \not{2}\ 13} \\ \not{8}\ \not{3}\ \not{3} \\ -\ 6\ 3\ 7 \\ \hline 1\ 9\ 6 \end{array}$$

Solution **The second book has 196 more pages than the first book.**

You can use mental math to subtract a **multiple** of 10, 100, or 1,000.

Here are some examples.

When you subtract 750 − 10, only the digit in the tens place of 750 will change. It will decrease by 1.

750 − 10 (or 1 ten) = 740

When you subtract 750 − 200, only the digit in the hundreds place of 750 will change. It will decrease by 2.

750 − 200 (or 2 hundreds) = 550

Example 3

What is 900 − 300?

Strategy **Use mental math.**

Step 1 Think: How many hundreds in 300?

There are 3 hundreds in 300.

Step 2 Think: Which digit will change in 900 when I subtract 3 hundreds?

9 is the digit in the hundreds place. It will decrease by 3.

Step 3 Subtract mentally.

900 − 300 (or 3 hundreds) = 600

Solution **900 − 300 = 600**

Example 4

Emily's family drove 847 miles on their camping trip. Trevor's family drove 100 miles less than Emily's family. How many miles did Trevor's family drive?

Strategy **Use mental math.**

Step 1 Write a subtraction number sentence for the problem.

Use the symbol ☐ for the difference.

847 − 100 = ☐

Step 2 Think: How many hundreds in 100?

There is 1 hundred in 100.

Step 3 Think: Which digit will change in 847 when I subtract 1 hundred?

The digit in the hundreds place is 8. It will decrease by 1.

Step 4 Subtract mentally.

847 − 100 (or 1 hundred) = 747

Solution **Trevor's family drove 747 miles.**

Coached Example

Diane counted 245 beads into a bowl. Zoe counted 10 less than Diane. How many beads did Zoe count?

Write a subtraction number sentence for the problem.

__________ − __________ = ☐

Use mental math.

How many tens in 10? ________

Which digit will change in 245 when you subtract 1 ten? __________

The digit in the tens place is ________. It will decrease by ________.

Mentally subtract.

Zoe counted __________ beads.

Lesson Practice

Choose the correct answer.

1. Neil has 373 CDs. Marcia has 229 CDs. How many more CDs does Neil have than Marcia?
   - A. 144
   - B. 145
   - C. 154
   - D. 155

2. Find the difference.

   474 − 392 = □
   - A. 82
   - B. 92
   - C. 122
   - D. 766

3. José had 250 raffle tickets to sell. There are 20 tickets left. How many tickets has José sold?
   - A. 230
   - B. 240
   - C. 270
   - D. 450

4. A movie theater sold 692 tickets to the evening show. It sold 385 tickets to the afternoon show. How many more tickets were sold for the evening show than the afternoon show?
   - A. 207
   - B. 217
   - C. 307
   - D. 317

5. What is 514 − 100?
   - A. 314
   - B. 414
   - C. 614
   - D. 714

6. A town has two schools. There are 473 students at Smith School. Washington School has 354 students. How many more students are at Smith School than at Washington School?
   - A. 119
   - B. 127
   - C. 129
   - D. 827

7. Henry has 570 baseball cards. Tom has 40 less baseball cards than Henry. How many baseball cards does Tom have?

- A. 170
- B. 470
- C. 520
- D. 530

8. Find the difference.

$301 - 148 = \square$

- A. 153
- B. 162
- C. 262
- D. 449

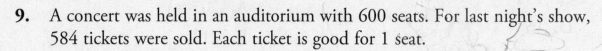

9. A concert was held in an auditorium with 600 seats. For last night's show, 584 tickets were sold. Each ticket is good for 1 seat.

A. How many tickets were **not** sold?

____

B. There were 26 people with tickets who did **not** make it to last night's show. How many empty seats were in the auditorium last night? Show your work.

Common Core State Standard:
3.NBT.1

Round Whole Numbers

Getting the Idea

You can **round** a number to the nearest ten or hundred.

Example 1

The dance recital was 73 minutes long. To the nearest ten minutes, about how long was the dance recital?

Strategy **Use a number line. Round to the nearest ten.**

Step 1 Make a number line from 70 to 80.

Find 73 on the number line.

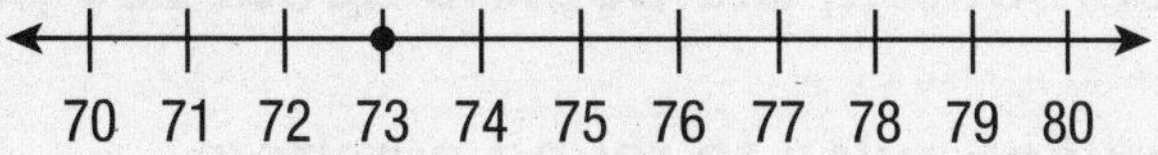

Step 2 Decide if 73 is closer to 70 or to 80.

73 is closer to 70 than to 80.

Round 73 down to 70.

Solution **To the nearest ten, the dance recital was about 70 minutes long.**

Example 2

John is reading a book with 247 pages. To the nearest ten, about how many pages long is the book?

Strategy **Use a number line to help you round to the nearest ten.**

Step 1 Make a number line from 240 to 250.

Find 247 on the number line.

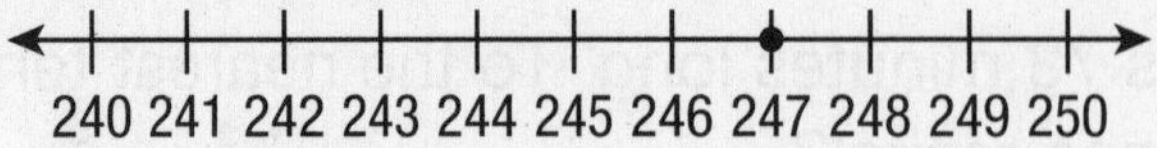

Step 2 Decide if 247 is closer to 240 or 250.

247 is closer to 250.

Round 247 up to 250.

Solution **To the nearest ten, the book is about 250 pages long.**

You can also use **rounding rules** to round numbers.
When you use rounding rules, look at the place to the right of the place you are rounding to.

- If the digit is less than 5 (1, 2, 3, or 4), round down.
- If the digit is 5 or greater (5, 6, 7, 8, or 9), round up.

Example 3

A basket has 193 apples. To the nearest hundred, about how many apples are in the basket?

Strategy **Use rounding rules to round to the nearest hundred.**

Step 1 The digit in the hundreds place is 1.

The digit to the right of the hundreds place is 9.

193

Step 2 Use rounding rules.

9 > 5, so round up.

193 rounded to the nearest hundred is 200.

Solution **To the nearest hundred, there are about 200 apples in the basket.**

Coached Example

Which is the greater number?

742 rounded to the nearest hundred
718 rounded to the nearest ten

Round 742 to the nearest hundred.

The digit in the hundreds place is ________.

The digit to the right of the rounding place is ________.

4 ◯ 5

Should you round 742 up or down? ____________

To the nearest hundred, 742 rounds to ________.

Round 718 to the nearest ten.

The digit in the tens place is ________.

The digit to the right of the rounding place is ________.

8 ◯ 5

Should you round 718 up or down? ____________

To the nearest ten, 718 rounds to ________.

Compare the rounded numbers.

________ < ________

________ **rounded to the nearest** ________ **is the greater number.**

Lesson Practice

Choose the correct answer.

1. Which shows 67 rounded to the nearest ten?

   ○ **A.** 60
   ○ **B.** 65
   ○ **C.** 70
   ○ **D.** 100

2. A store has 38 items on sale this week. To the nearest ten, about how many items are on sale at the store?

   ○ **A.** 40
   ○ **B.** 38
   ○ **C.** 35
   ○ **D.** 30

3. An amusement park had 276 visitors one day. To the nearest ten, about how many visitors did the park have that day?

   ○ **A.** 200
   ○ **B.** 270
   ○ **C.** 280
   ○ **D.** 300

4. Which shows 112 rounded to the nearest hundred?

   ○ **A.** 100
   ○ **B.** 110
   ○ **C.** 120
   ○ **D.** 130

5. Ms. Cannon bowled a 185. What was her score to the nearest hundred?

   ○ **A.** 100
   ○ **B.** 190
   ○ **C.** 200
   ○ **D.** 300

6. Anthony collected 778 signatures for a petition. To the nearest hundred, about how many signatures did he collect?

   ○ **A.** 700
   ○ **B.** 780
   ○ **C.** 800
   ○ **D.** 1,000

7. Which is the greatest number?

○ **A.** 503 rounded to the nearest ten

○ **B.** 474 rounded to the nearest hundred

○ **C.** 585 rounded to the nearest ten

○ **D.** 570 rounded to the nearest hundred

8. To the nearest hundred, which number does **not** round to 400?

○ **A.** 348

○ **B.** 372

○ **C.** 419

○ **D.** 446

9. A candy store is running a promotion where customers can guess the number of jelly beans in a jar and win a prize. If customers guess the number to the nearest 10 or 100, they also receive smaller prizes. There are 838 jelly beans in the jar.

A. To the nearest ten, how many jelly beans are in the jar? 840

B. To the nearest hundred, how many jelly beans are in the jar? 900

Domain 1 • Lesson 8

Common Core State Standard:
3.OA.8

Estimate Sums and Differences

Getting the Idea

You can **estimate** sums or differences in problems. An estimate is a number close to the exact answer.

Example 1

Estimate the sum.

42 + 38 + 54

Strategy **Round each number to the greatest place. Then add.**

Step 1 Find the greatest place of each number.

The greatest place of each number is the tens place.

Step 2 Round each number to the nearest ten.

42 rounds down to 40 because $2 < 5$.

38 rounds up to 40 because $8 > 5$.

54 rounds down to 50 because $4 < 5$.

Step 3 Add the rounded numbers.

$40 + 40 + 50 = 130$

Solution **The estimated sum of 42 + 38 + 54 is 130.**

Example 2

Polk Elementary School has 413 students. Grant Elementary School has 278 students. About how many more students attend Polk Elementary School than Grant Elementary School?

Strategy **Decide if you need an exact answer or an estimate. Then solve.**

Step 1 Decide if you need an exact answer or an estimate.

The problem asks "about how many," so estimate.

Step 2 Round each number to the greatest place.

The greatest place of each number is the hundreds place.

413 rounds down to 400 because $1 < 5$.

278 rounds up to 300 because $7 > 5$.

Step 3 Decide if you need to add or subtract.

The problem asks "about how many more," so subtract.

$400 - 300 = 100$

Solution **About 100 more students attend Polk Elementary School than Grant Elementary School.**

You can use estimation to check if an answer is reasonable, or if the answer makes sense.

Example 3

Molly has $87 in her wallet. She bought a pair of pants for $32. She said she has about $30 left. Is Molly's answer reasonable?

Strategy **Round each number to the greatest place. Then subtract.**

Step 1 Decide how to solve the problem.

Estimate to see if Molly's answer is reasonable.

"She has about $30 left" tells you to subtract.

Estimate the difference of $87 − $32.

Step 2 Round each amount to the nearest ten dollars. Then subtract.

$87 rounds up to $90. $32 rounds down to $30.

$90 − $30 = $60

Step 3 Compare the difference to Molly's answer.

$60 is not close to $30.

Solution **Molly's answer is not reasonable.**

When you solve a problem, it is helpful to estimate the answer before you find the exact answer. Then use the estimate to check if the exact answer is reasonable.

Example 4

Yesterday, a museum had 718 visitors. Today, the museum had 95 more visitors than yesterday. How many visitors did the museum have today?

Strategy **Find an estimate first.**
Then compare the estimate to the exact answer.

Step 1 Decide how to solve the problem.

"95 more visitors than yesterday" tells you to add.

Step 2 Estimate the sum of 718 + 95.

718 rounds down to 700.

95 rounds up to 100.

700 + 100 = 800

The answer should be about 800.

Step 3 Find the exact sum of 718 + 95.

$$\begin{array}{r} {\scriptstyle 11} \\ 718 \\ +\ 95 \\ \hline 813 \end{array}$$

Step 4 Compare the exact answer to the estimate.

813 is close to 800.

The answer is reasonable.

Solution **The museum had 813 visitors today.**

Coached Example

Mr. Mitchell bought a computer that cost $482. He also bought a printer that cost $117. How much did Mr. Mitchell spend in all?

Decide how to solve the problem.

"How much did Mr. Mitchell spend in all?" tells you to ____________.

Estimate the sum of $__________ + $__________.

Round each number to the greatest place.

$482 rounds up to $__________.

$117 rounds down to $__________.

Add the rounded amounts.

$__________ + $__________ = $__________

The answer should be about $__________.

Find the exact sum.

Compare the exact answer to the estimate.

Is the exact amount close to the estimate? __________

Is your exact answer reasonable? __________

Mr. Mitchell spent $__________ in all.

Lesson Practice

Choose the correct answer.

1. Which shows the best way to estimate the sum?

   82 + 59 + 25

   - A. 80 + 50 + 30
   - B. 60 + 60 + 60
   - C. 70 + 50 + 20
   - D. 80 + 60 + 30

2. Which is the best estimate of the sum?

   184 + 245

   - A. 300
   - B. 400
   - C. 600
   - D. 750

3. Which shows the best way to estimate the difference?

   613 − 178

   - A. 500 − 200
   - B. 600 − 100
   - C. 500 − 200
   - D. 600 − 200

4. Which is the best estimate of the difference?

   404 − 159

   - A. 50
   - B. 100
   - C. 200
   - D. 400

5. A farmer planted 83 tomato seeds and 76 pumpkin seeds. Which shows the best estimate of the number of seeds the farmer planted in all?

   - A. 160
   - B. 200
   - C. 260
   - D. 300

6. Ms. Jenkins spent $94 on a chair and $185 on a table. About how much more did the table cost than the chair?

   - A. $50
   - B. $100
   - C. $150
   - D. $200

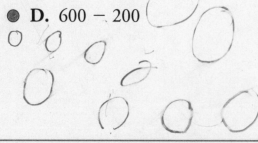

7. John is reading a book with 332 pages. He has read 128 pages. Before you find the exact number of pages John has left to read, which can use you use to check if the answer is reasonable?

- ○ **A.** 300 + 100
- ○ **B.** 300 − 200
- ○ **C.** 300 − 100
- ○ **D.** 400 − 100

8. An amusement park had 376 visitors in the morning. The park had another 145 visitors in the evening. About how many visitors did the park have that day?

- ○ **A.** 200
- ○ **B.** 300
- ○ **C.** 400
- ○ **D.** 500

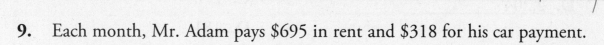

9. Each month, Mr. Adam pays $695 in rent and $318 for his car payment.

A. About how much more is his rent than his car payment? Show your work.

B. Exactly how much more is Mr. Adam's rent than his car payment? Explain how you know your answer is reasonable.

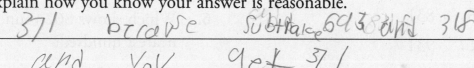

371 brcave subtrake 695 and 318 and you get 371

Domain 1: Cumulative Assessment for Lessons 1–8

1. Isa made a chain that is 1,204 paper clips long. Rosita made a chain that is 100 paper clips longer. How many paper clips are in Rosita's chain?

   ○ **A.** 1,104
   ○ **B.** 1,304
   ○ **C.** 1,314
   ○ **D.** 2,204

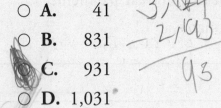

2. Lincoln Elementary sold 3,124 raffle tickets for the fair. Union Elementary sold 2,193 raffle tickets. How many more raffle tickets did Lincoln Elementary sell than Union Elementary?

   ○ **A.** 41
   ○ **B.** 831
   ○ **C.** 931
   ○ **D.** 1,031

3. Which of the following is true?

   ○ **A.** 1,274 = 1,742
   ○ **B.** 2,951 > 2,995
   ○ **C.** 1,803 < 1,903
   ○ **D.** 2,273 < 2,172

4. Lana and Ernie built two block towers. The first tower had 281 blocks. The second tower had 100 more blocks than the first tower. How many blocks were in the second tower?

   ○ **A.** 381
   ○ **B.** 271
   ○ **C.** 270
   ○ **D.** 181

5. What is 4,108 − 1,000?

   ○ **A.** 5,108
   ○ **B.** 4,008
   ○ **C.** 3,108
   ○ **D.** 3,107

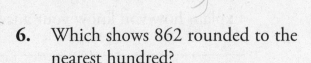

6. Which shows 862 rounded to the nearest hundred?

   ○ **A.** 800
   ○ **B.** 860
   ○ **C.** 870
   ○ **D.** 900

7. Which is the missing number?

(37 + □) + 25 = 37 + (45 + 25)

- ○ **A.** 25
- ○ **B.** 37
- ○ **C.** 45
- ○ **D.** 107

8. Jack said that a good estimate for 691 − 476 is 200. Julio said it is 210. Who is correct?

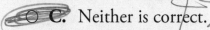

- ○ **A.** Jack is correct.
- ○ **B.** Julio is correct.
- ○ **C.** Neither is correct.
- ○ **D.** Both are correct.

9. What is the missing number in this pattern?

32 28 24 ? 16 12

____

10. Thomson Elementary School has 490 students. Redding Elementary School has 385 students.

A. How many students do the schools have in all?

____

B. Weston Elementary School has 100 more students than Thomson Elementary School. How many students does Weston Elementary School have?

____

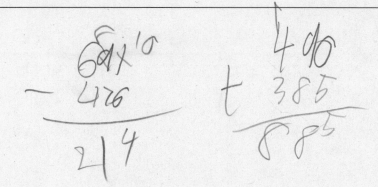

Domain 2

Operations and Algebraic Thinking

Domain 2: Diagnostic Assessment for Lessons 9–18

Domain 2: Cumulative Assessment for Lessons 9–18

Domain 2: Diagnostic Assessment for Lessons 9–18

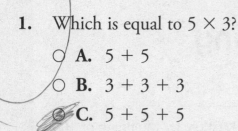

1. Which is equal to 5×3?

- ○ **A.** $5 + 5$
- ○ **B.** $3 + 3 + 3$
- ○ **C.** $5 + 5 + 5$
- ○ **D.** $5 + 3 + 5 + 3$

2. Which shows the commutative property of multiplication?

- ○ **A.** $8 \times 6 = 6 \times 8$
- ○ **B.** $6 \times 1 = 6$
- ○ **C.** $8 \times 6 = 8 \times (3 + 3)$
- ○ **D.** $8 + 6 = 6 + 8$

3. Find the product.

$4 \times 7 = \square$

- ○ **A.** 11
- ○ **B.** 14
- ○ **C.** 21
- ○ **D.** 28

4. Which number makes both sentences true?

$24 \div \square = 6$

$6 \times \square = 24$

- ○ **A.** 30
- ○ **B.** 8
- ○ **C.** 6
- ○ **D.** 4

5. Connor has 7 bags of marbles. Each bag has 8 marbles in it. How many marbles does Connor have in all?

- ○ **A.** 56
- ○ **B.** 49
- ○ **C.** 42
- ○ **D.** 15

6. The table shows the total number of dumplings for different numbers of orders.

Dumplings Ordered

Number of Orders	Number of Dumplings
2	16
4	32
6	48
8	64

How many dumplings are in 10 orders?

- ○ **A.** 65
- ○ **B.** 72
- ○ **C.** 80
- ○ **D.** 88

7. Which division fact does this picture show?

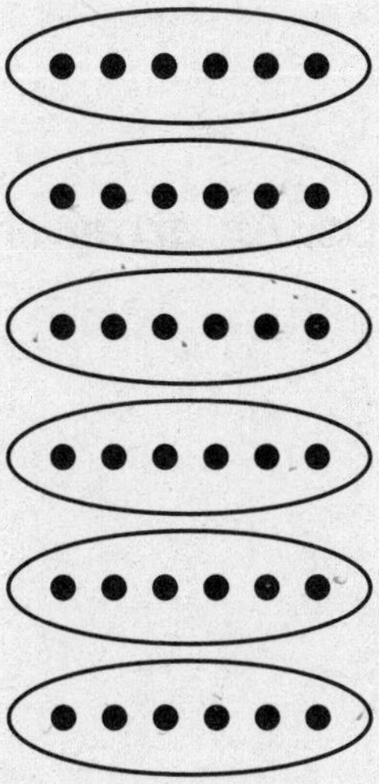

- ○ **A.** 36 ÷ 6 = 6
- ○ **B.** 36 ÷ 9 = 4
- ○ **C.** 36 ÷ 4 = 9
- ○ **D.** 40 ÷ 5 = 8

8. Which multiplication fact can be used to find the missing number?

35 ÷ □ = 7

- ○ **A.** 7 × 1 = 7
- ○ **B.** 7 × 4 = 28
- ○ **C.** 7 × 5 = 35
- ○ **D.** 7 × 7 = 49

9. There are 10 tea candles in a box. Mrs. Sullivan bought 7 boxes. How many tea candles did she buy?

70 candles

10. Rhonda has 3 bunches of flowers. Each bunch has 10 flowers.

A. Draw a model of the problem.

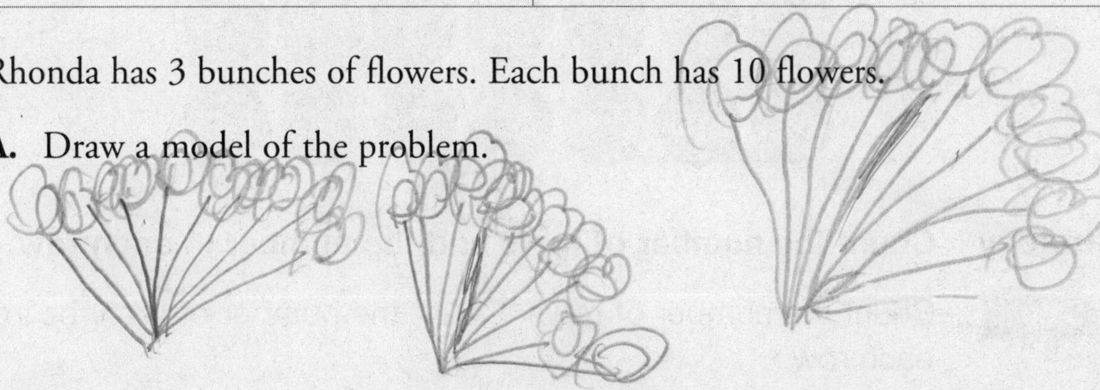

B. Write a multiplication sentence for the problem. Use the symbol □ for the product.

3 × 10 = □ 30

C. How many flowers does Rhonda have?

Rhonda has 30 flowers

Domain 2 • Lesson 9

Understand Multiplication

Common Core State Standards: 3.OA.1, 3.OA.3, 3.OA.4

Getting the Idea

You can use **multiplication** to combine equal groups. An **array** shows equal groups of objects in rows and columns.

Here are the parts of a multiplication sentence.

3	×	2	=	6
factor		**factor**		**product**

Example 1

Write a multiplication sentence for this array.

Strategy **Count the number of rows and the number in each row.**

Step 1 Count the number of rows. Count the number of oranges in each row.

There are 4 rows.

Each row has 7 oranges.

Step 2 Find the total number of oranges.

4 groups of 7 equals 28.

Step 3 Write the multiplication sentence.

$4 \times 7 = 28$

Solution **The array shows the multiplication sentence $4 \times 7 = 28$.**

Repeated addition is adding the same number many times.

Repeated addition and multiplication have the same result.

You can use repeated addition to solve a multiplication problem.

Example 2

How many stars are there in all?

$3 \times 4 = \square$

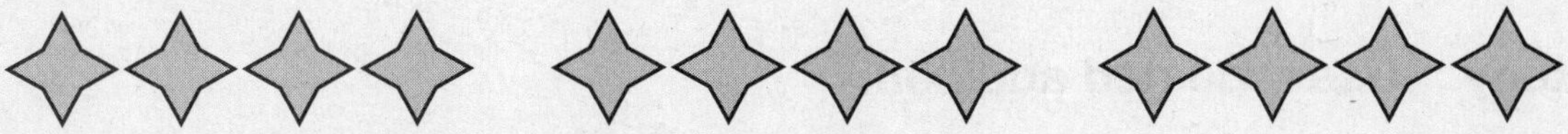

Strategy **Use repeated addition.**

Step 1 Count the number of stars in each group. Count the number of equal groups.

There are 4 stars in each group.

There are 3 equal groups.

Step 2 Use repeated addition.

Add 4 three times to find the total.

$4 + 4 + 4 = 12$

Solution **There are 12 stars in all.**

$\mathbf{3 \times 4 = 12}$

Example 3

How many pencils are there in all?

Write an addition sentence and a multiplication sentence.

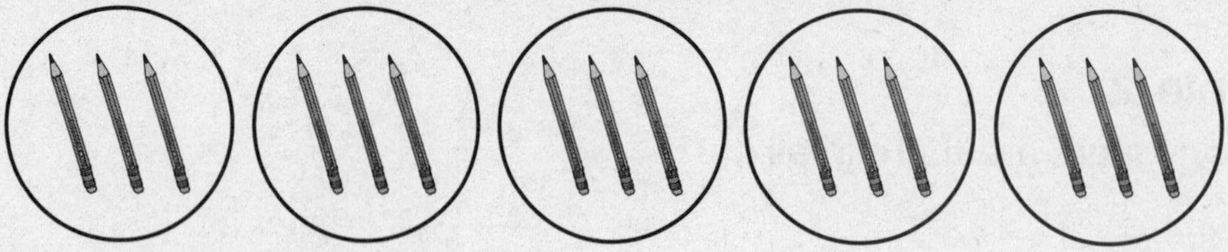

Strategy **Use repeated addition.**

Step 1 Count the number of pencils in each group. Count the number of equal groups.

There are 3 pencils in each group.

There are 5 equal groups.

Step 2 Write an addition sentence.

Add 3 five times to find the total.

$3 + 3 + 3 + 3 + 3 = 15$

Step 3 Write a multiplication sentence.

5 groups of 3 equals 15.

$5 \times 3 = 15$

Solution **There are 15 pencils in all.**

$\mathbf{3 + 3 + 3 + 3 + 3 = 15}$

$\mathbf{5 \times 3 = 15}$

You can use a rectangular area model to show multiplication.

Example 4

What multiplication sentence does this rectangular area model show?

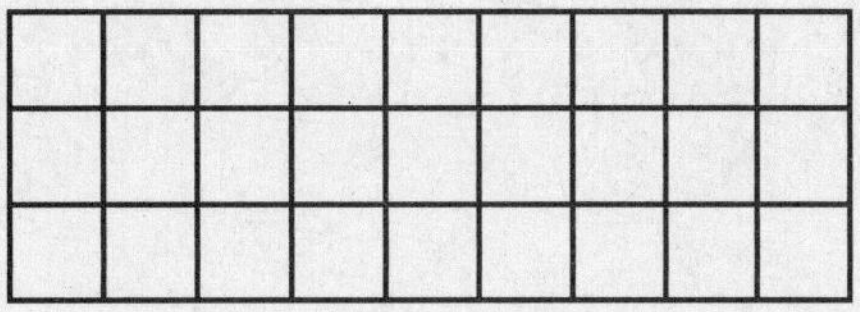

Strategy **Use an area model to multiply.**

Step 1 Count the number of rows and columns.

There are 3 rows and 9 columns.

There are 27 squares in all.

Step 2 Write a multiplication sentence.

$3 \times 9 = 27$ or $9 \times 3 = 27$

Solution **The rectangular area model shows $3 \times 9 = 27$ and $9 \times 3 = 27$.**

Coached Example

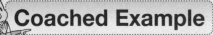

Reiko put 2 cookies on each plate.

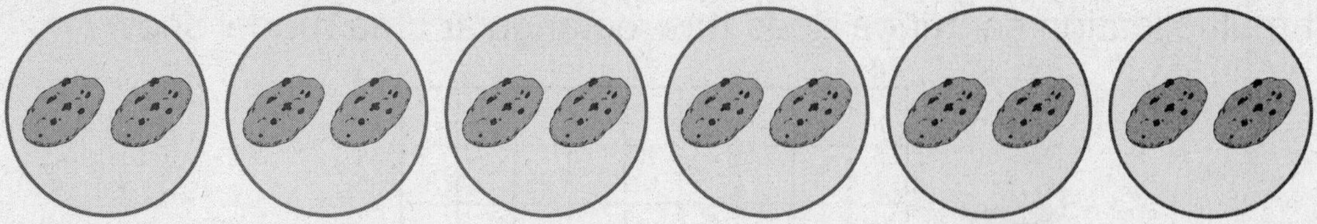

How many cookies are there in all?

Count the number of cookies on each plate.

There are ________ cookies on each plate.

Count the number of plates.

There are ________ plates.

Write an addition sentence.

________ + ________ + ________ + ________ + ________ + ________ = ________

Write a multiplication sentence.

________ groups of ________ equals ________.

________ × ________ = ________

There are ________ cookies in all.

Lesson Practice

Choose the correct answer.

1. Which multiplication sentence does this array show?

○ **A.** $3 \times 2 = 6$

○ **B.** $5 \times 2 = 10$

○ **C.** $3 \times 5 = 15$

○ **D.** $5 \times 5 = 25$

2. Which multiplication sentence does this array show?

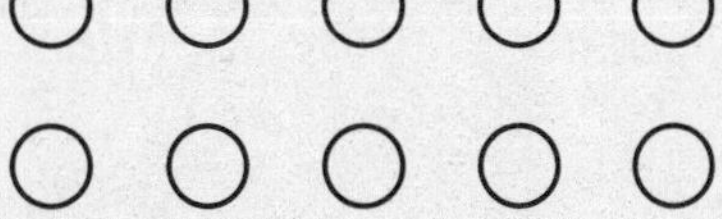

○ **A.** $5 \times 5 = 25$

○ **B.** $2 \times 6 = 12$

○ **C.** $2 \times 5 = 10$

○ **D.** $2 + 5 = 7$

3. Which multiplication sentence shows the total number of children?

○ **A.** $4 \times 6 = 24$

○ **B.** $3 \times 7 = 21$

○ **C.** $3 \times 4 = 12$

○ **D.** $7 \times 1 = 7$

4. Which addition sentence does this picture show?

○ **A.** $7 + 7 + 7 + 7 = 28$

○ **B.** $4 + 4 + 4 + 4 = 16$

○ **C.** $4 + 7 + 4 + 7 = 24$

○ **D.** $4 + 4 + 4 = 12$

5. Which multiplication sentence does this area model show?

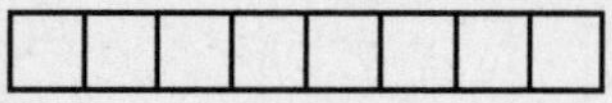

- ○ **A.** $2 \times 4 = 8$
- ◉ **B.** $1 \times 8 = 8$
- ○ **C.** $8 \times 8 = 64$
- ○ **D.** $7 \times 2 = 14$

6. Which is equal to 6×4?

- ○ **A.** $4 + 4 + 4 + 4$
- ◉ **B.** $6 + 6 + 6 + 6$
- ○ **C.** $6 + 6 + 6 + 6 + 6 + 6$
- ○ **D.** $6 + 4 + 6 + 4$

7. Which area model shows $5 \times 5 = 25$?

○ **A.**

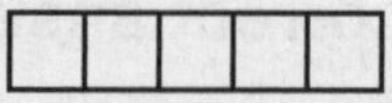

○ **B.**

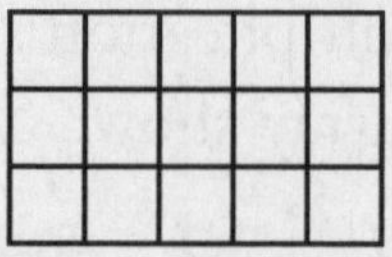

○ **C.**

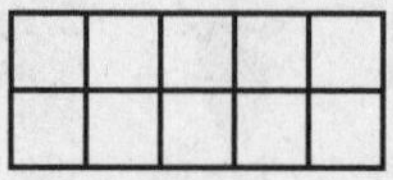

◉ **D.**

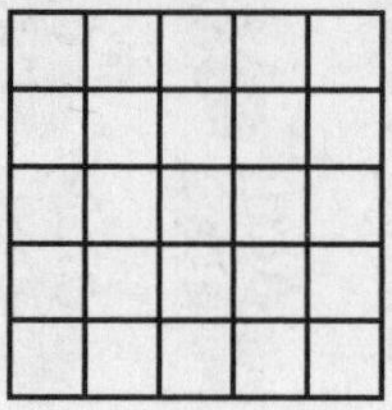

8. Look at the equal groups below.

A. Write an addition sentence to show how many stars in all.

____

B. Write a multiplication sentence to show how many stars in all.

____

Multiplication Facts

Getting the Idea

There are many strategies you can use to solve multiplication problems.

For example, to find the product of $6 \times 6 = \square$:

You can use skip counting.

6, 12, 18, 24, 30, 36

You can use repeated addition.

$6 + 6 + 6 + 6 + 6 + 6 = 36$

You can use a multiplication table.

The factors are along the top row and down the first column on the left. The products fill out the rest of the table.

So, $6 \times 6 = 36$.

Factors (top row); **Factors** (left column)

×	0	1	2	3	4	5	6	7	8	9	10
0	0	0	0	0	0	0	0	0	0	0	0
1	0	1	2	3	4	5	6	7	8	9	10
2	0	2	4	6	8	10	12	14	16	18	20
3	0	3	6	9	12	15	18	21	24	27	30
4	0	4	8	12	16	20	24	28	32	36	40
5	0	5	10	15	20	25	30	35	40	45	50
6	0	6	12	18	24	30	**36**	42	48	54	60
7	0	7	14	21	28	35	42	49	56	63	70
8	0	8	16	24	32	40	48	56	64	72	80
9	0	9	18	27	36	45	54	63	72	81	90
10	0	10	20	30	40	50	60	70	80	90	100

You can use the multiplication table to find the product of 0 and a factor.

Look at the products for 0 in the multiplication table.

Notice that any number times 0 is equal to 0.

For example, $3 \times 0 = 0$.

You can also use the multiplication table to find the product of 1 and a factor.

Look at the products for 1 in the multiplication table.

Notice that any number times 1 is that number.

For example, $9 \times 1 = 9$.

Example 1

Find the product.

$5 \times 10 = \square$

Strategy **Use a multiplication table.**

Look at the 5s row. Find the 10s column.

Now, find the box where the row and the column meet.

The number inside the box, 50, is the product.

Solution **$5 \times 10 = 50$**

When you multiply any whole number by 10, the product is the whole number with a zero written in the ones place.

For example, $5 \times 10 = 50$.

Example 2

Find the product.

$4 \times 2 = \square$

Strategy **Use skip counting.**

Use a number line to skip count by 2s four times.

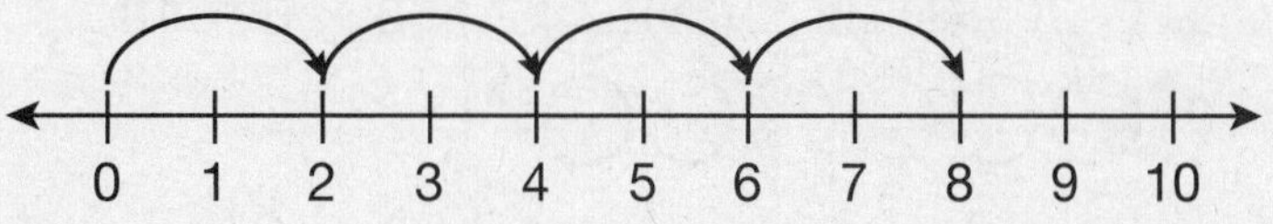

Solution **$4 \times 2 = 8$**

When you multiply by 2, you can use doubling to find the product.

Example 3

How many cookies are there in all?

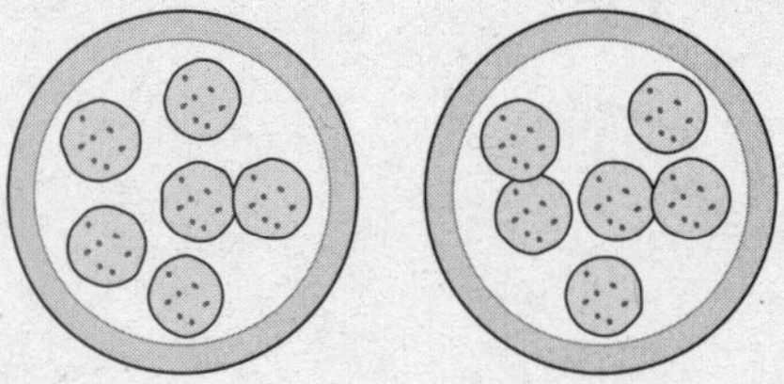

Strategy **Use doubling.**

Step 1 Look at the picture.

There are 2 plates.

Each plate has 6 cookies.

Step 2 Double 6.

$6 + 6 = 12$

Step 3 Write a multiplication sentence.

2 groups of 6 equals 12.

$2 \times 6 = 12$

Solution **There are 12 cookies in all.**

$\mathbf{2 \times 6 = 12}$

You can double a multiplication fact you already know to find a new fact.

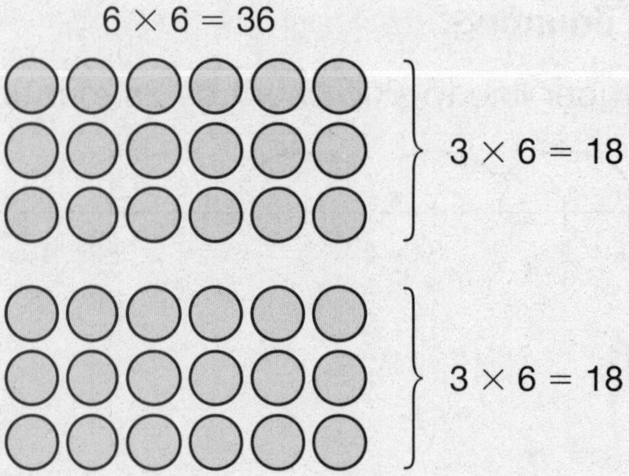

Example 4

Find the product.

$8 \times 7 = \square$

Strategy **Double a known fact.**

Step 1 One of the factors is 8.

8 is a double of 4.

Step 2 Think of a known fact: 4×7.

$4 \times 7 = 28$

Step 3 8 is the double of 4, so double the product of 4×7.

$28 + 28 = 56$, so $8 \times 7 = 56$.

Solution $\mathbf{8 \times 7 = 56}$

You can find a missing factor or product in a multiplication problem using a variety of strategies. A missing number can be represented by a box ($\square$) or a letter (*x*).

Example 5

Find the missing factor.

$\square \times 3 = 15$

Strategy **Use skip counting.**

Step 1 Use a number line to skip count by 3s.

Skip count by 3 until you reach 15.

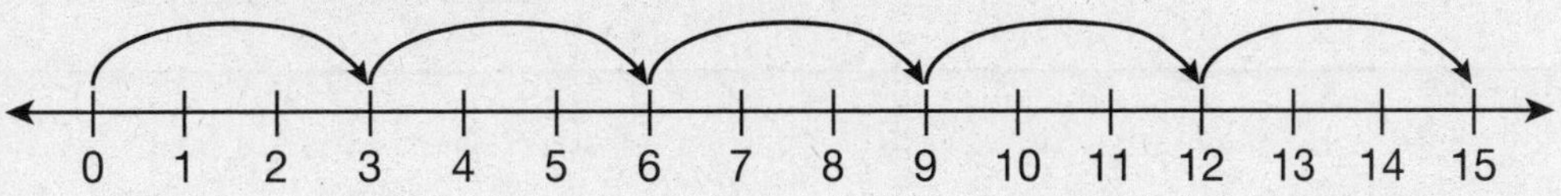

Step 2 Count the number of times you skip counted.

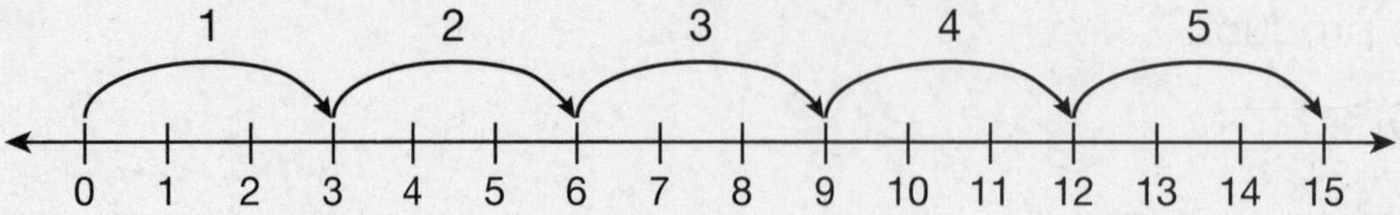

You skip counted 5 times.

5 is the missing factor.

Solution $5 \times 3 = 15$

Coached Example

Write a multiplication sentence for this model.

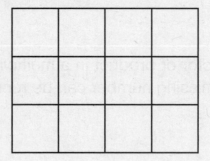

How many rows are there? ______

How many squares are in each row? ______

Use skip counting to find the total number of squares.

______, ______, ______

The model shows the multiplication sentence ______ × ______ = ______.

Lesson Practice

Choose the correct answer.

1. Find the product.

$3 \times 3 = \square$

- ○ **A.** 6
- ○ **B.** 9
- ○ **C.** 11
- ○ **D.** 30

2. Find the product.

$2 \times 9 = \square$

- ○ **A.** 11
- ○ **B.** 18
- ○ **C.** 29
- ○ **D.** 92

3. Find the missing factor.

$6 \times \square = 42$

- ○ **A.** 4
- ○ **B.** 5
- ○ **C.** 6
- ○ **D.** 7

4. Find the product.

$6 \times 10 = \square$

- ○ **A.** 6
- ○ **B.** 10
- ○ **C.** 16
- ○ **D.** 60

5. Find the product.

$4 \times 5 = \square$

- ○ **A.** 9
- ○ **B.** 15
- ○ **C.** 20
- ○ **D.** 45

6. Find the product.

$8 \times 8 = \square$

- ○ **A.** 88
- ○ **B.** 64
- ○ **C.** 18
- ○ **D.** 16

7. Find the product.

$9 \times 5 = \square$

○ **A.** 14
○ **B.** 35
○ **C.** 40
○ **D.** 45

8. Find the missing factor.

$\square \times 2 = 14$

○ **A.** 12
○ **B.** 7
○ **C.** 4
○ **D.** 2

9. Look at the array below.

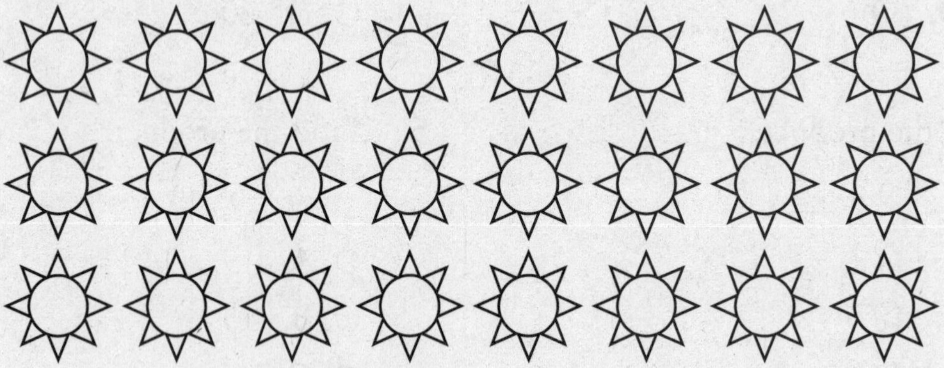

A. Skip count to find how many in all.

____________, ____________, ____________

B. Write a multiplication fact for this array.

____________ × ____________ = ____________

Common Core State Standard:
3.OA.9

Multiplication Patterns

Getting the Idea

If you add the same number to itself over and over, you are creating a pattern. Since multiplication is another way to do repeated addition, you can use multiplication as a shortcut.

Example 1

There are 5 pieces of paper in each pile.

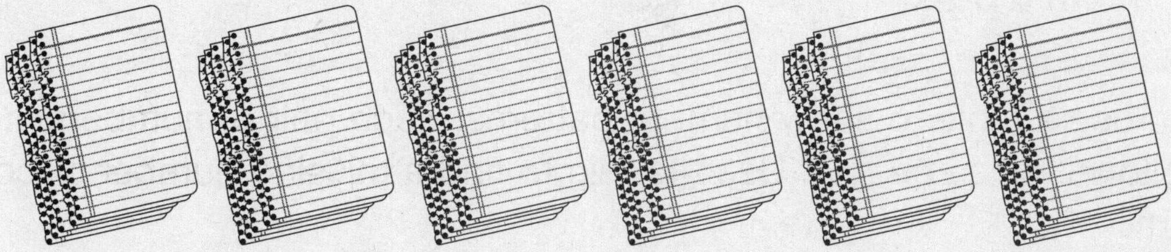

Write a multiplication sentence to show the total number of pieces in 6 piles.

Strategy **Relate repeated addition to multiplication.**

Step 1 Find the number of pieces in 6 piles.

There are 5 pieces in each pile. The rule is add 5.

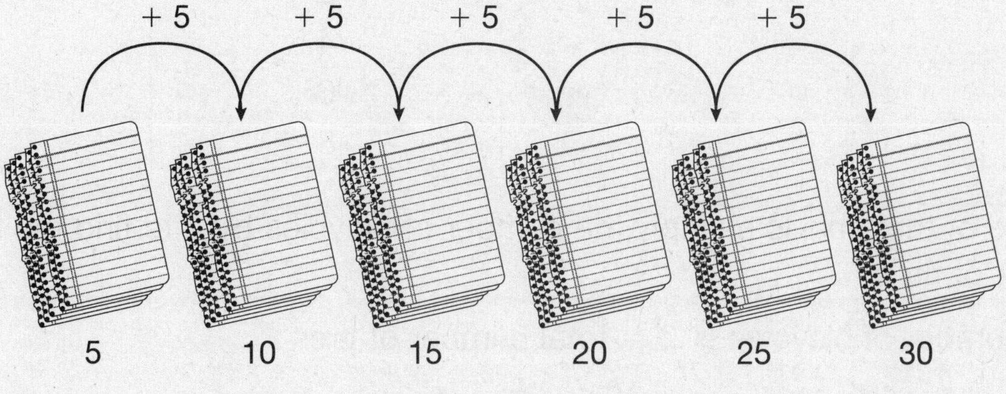

Step 2 Use multiplication to show the total number of pieces.

number of piles × pieces in each pile = total

$1 \times 5 = 5$

$2 \times 5 = 10$

$3 \times 5 = 15$

$4 \times 5 = 20$

$5 \times 5 = 25$

$6 \times 5 = 30$

Solution **The sentence $6 \times 5 = 30$ shows the total number of pieces in 6 piles.**

You can use a table to show number patterns. Each pair of numbers in the table follows the same rule. Use the rule to find a missing number or continue a pattern.

For example, the table below shows that 1 bicycle has 2 tires, 2 bicycles have a total of 4 tires, 3 bicycles have a total of 6 tires, and 4 bicycles have a total of 8 tires. How many tires do 5 bicycles have?

Bicycle Tires

Number of Bicycles	Number of Tires
1	2
2	4
3	6
4	8

The rule of the table is multiply the number of bicycles by 2 to find the total number of tires.

Rule: number of bicycles × 2 = total number of tires

Use the rule to find how many tires 5 bicycles have.

$5 \times 2 = 10$

So, 5 bicycles have 10 tires.

Example 2

This table shows the total amount of milk needed to make different numbers of cakes.

Cups of Milk Needed

Number of Cakes	Cups of Milk
1	3
2	6
3	9
4	12

How many cups of milk in all are needed to make 5 cakes?

Strategy **Find the rule.**

Step 1 Find the number of cups of milk for 1 cake.

3 cups of milk are needed for 1 cake.

Step 2 Look at the pattern for the rest of the numbers.

As the number of cakes increases by 1, the cups of milk increase by 3.

Step 3 Find the rule.

The rule is number of cakes × 3 cups = total cups of milk.

Step 4 Use the rule to find the total cups of milk for 5 cakes.

5 cakes × 3 cups = 15 cups

Solution **15 cups of milk are needed to make 5 cakes.**

There are many patterns in a multiplication table.

For example, look at the products in the 4 row.

The numbers increase by 4 as you go from left to right.

×	1	2	3	4	5	6	7	8	9	10
1	**1**	2	3	4	5	6	7	8	9	10
2	2	**4**	6	8	10	12	14	16	18	20
3	3	6	**9**	12	15	18	21	24	27	30
4	4	8	12	**16**	20	24	28	32	36	40
5	5	10	15	20	**25**	30	35	40	45	50
6	6	12	18	24	30	**36**	42	48	54	60
7	7	14	21	28	35	42	**49**	56	63	70
8	8	16	24	32	40	48	56	**64**	72	80
9	9	18	27	36	45	54	63	72	**81**	90
10	10	20	30	40	50	60	70	80	90	**100**

The boldfaced numbers are the products of two equal factors, such as $3 \times 3 = 9$. Notice the products as you move diagonally from top left to bottom right. The numbers alternate between odd and even numbers.

Notice also that the products on one side of the diagonal are the same as the products on the other side of the diagonal.

For example, $7 \times 2 = 14$ and $2 \times 7 = 14$.

Example 3

When you multiply a number by 6, the product will always be an even number.

Show the product of 6×3 as two equal addends.

Strategy **Think about the definition of an even number.**

Step 1 What is an even number?

An even number can be separated into 2 equal groups.

Step 2 Find the product of 6×3.

$6 \times 3 = 18$

Step 3 Show 18 as two equal addends.

"Two equal addends" means the addends are the same.

Think: $? + ? = 18$

$9 + 9 = 18$

Solution **The product of 6×3 can be shown as $9 + 9 = 18$.**

Coached Example

This table shows the total number of legs for different numbers of spiders.

Spider Legs

Number of Spiders	Number of Legs
1	8
3	24
5	40
7	56

How many legs in all do 9 spiders have?

Find the number of legs for 1 spider.

Each spider has __________ legs.

Find the rule.

The rule is number of spiders × __________ = total number of legs.

Use the rule to check the numbers in the table.

1 × __________ = 8

3 × __________ = 24

5 × __________ = 40

7 × __________ = 56

Use the rule to find the total number of legs that 9 spiders have.

9 × __________ = __________

There are __________ legs in all in 9 spiders.

Nine spiders have __________ legs in all.

Lesson Practice

Choose the correct answer.

1. There are 3 carrots in each bunch.

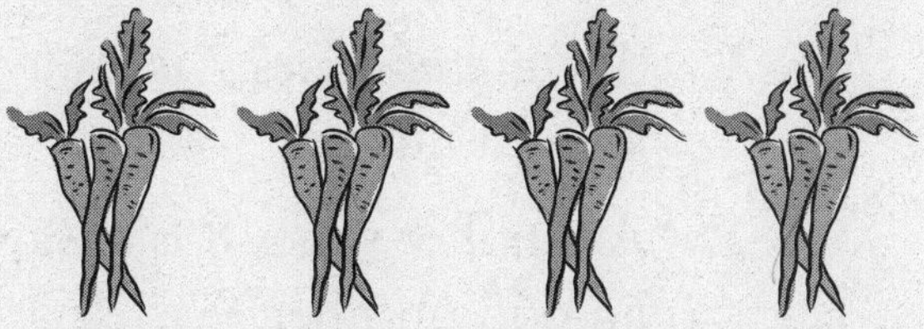

Which shows how many carrots are in 4 bunches?

- ○ **A.** 3 + 3 = 6
- ○ **B.** 4 × 2 = 8
- ○ **C.** 4 × 3 = 12
- ○ **D.** 8 × 2 = 16

2. Each muffin has 3 raisins on it.

How many raisins are on 5 muffins?

- ○ **A.** 10
- ○ **B.** 12
- ○ **C.** 15
- ○ **D.** 18

3. Which is the same as the product of 4 × 3?

- ○ **A.** 5 + 2
- ○ **B.** 7 + 7
- ○ **C.** 8 + 6
- ○ **D.** 6 + 6

Use the table below for questions 4 and 5.

Kate uses the same number of cups of pecans for each pie. The table shows the total cups of pecans for different numbers of pies.

Number of Pies	Cups of Pecans
1	3
2	6
3	9
4	12

4. What is the rule of the table?

- ○ **A.** number of pies × 3 = cups of pecans
- ○ **B.** number of pies + 3 = cups of pecans
- ○ **C.** number of pies × 6 = cups of pecans
- ○ **D.** number of pies + 9 = cups of pecans

5. How many cups of pecans are needed for 6 pies?

- ○ **A.** 14 cups
- ○ **B.** 15 cups
- ○ **C.** 17 cups
- ○ **D.** 18 cups

6. Samantha recorded the number of miles she ran in a table.

Miles Run

Number of Days	Number of Miles
2	4
4	8
6	12
8	16

If the pattern continues, how many miles will Samantha run in 10 days?

- **A.** 18 miles
- **B.** 20 miles
- **C.** 22 miles
- **D.** 24 miles

7. Which sentence is true?

- **A.** A number times 2 is always an even number.
- **B.** A number times 5 is always an even number.
- **C.** A number times 7 is always an odd number.
- **D.** A number times 8 is always an odd number.

8. Joann is making a quilt out of squares. Each square is made by sewing triangles together. The table shows how many triangles are needed to make different numbers of squares.

Number of Squares	Number of Triangles
2	4
4	8
6	12
8	16

A. How many triangles are needed to make 7 squares? Explain your answer.

B. Is the number of triangles needed for 7 squares odd or even? Explain your answer.

Common Core State Standards: 3.OA.3, 3.OA.8

Multiplication Word Problems

Getting the Idea

With multiplication word problems, look for key phrases to help you solve the problem.

How many in all means you need to find the product.

How many in each group or *how many groups* means you need to find one of the factors.

Remember, the numbers being multiplied are the factors and the answer is the product.

When you write a multiplication sentence, remember to use a symbol or letter to represent the unknown number.

Example 1

In a classroom, there are 3 rows of student desks. There are 6 student desks in each row.

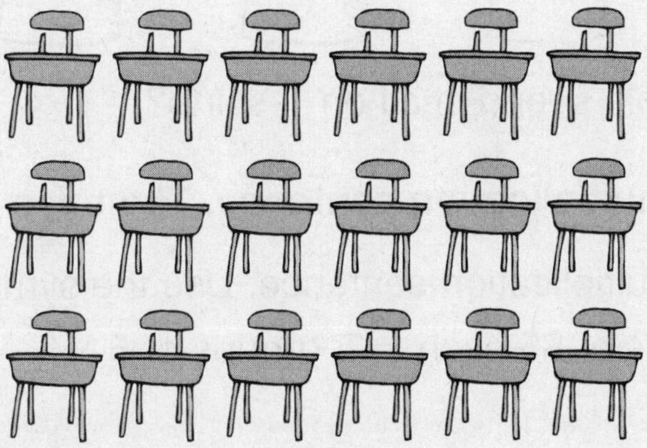

How many student desks are in the classroom?

Strategy **Write a multiplication sentence. Then double a known fact to solve.**

Step 1 Write a multiplication sentence.

There are 3 rows. There are 6 desks in each row.

number of rows × number in each row = total number

$3 \times 6 = \square$

Step 2 One of the factors is 6.

6 is a double of 3.

Step 3 Think of a known fact: 3×3.

$3 \times 3 = 9$

Step 4 6 is the double of 3, so double the product of 3×3.

$9 + 9 = 18$

So, $3 \times 6 = 18$.

Solution **There are 18 student desks in the classroom.**

Example 2

Mr. Cole bought 5 T-shirts. Each T-shirt costs $5.

How much did Mr. Cole spend in all on T-shirts?

Strategy **Write a multiplication sentence. Then use repeated addition.**

Step 1 Write a multiplication sentence. Use the symbol □ for the product.

5 shirts at \$5 each = 5 groups of 5

$5 \times 5 = \square$

Step 2 Add 5 five times.

$5 + 5 + 5 + 5 + 5 = 25$

So, $5 \times 5 = 25$.

Solution **Mr. Cole spent $25 on 5 T-shirts.**

Example 3

Three groups signed up to hike on a trail. Each group has 7 people. How many people in all are on the trail?

Strategy **Write a multiplication sentence. Then use skip counting to solve.**

Step 1 Write a multiplication sentence.

3 groups of 7 people = 3 groups of 7

$3 \times 7 = \square$

Step 2 Skip count by 7 three times.

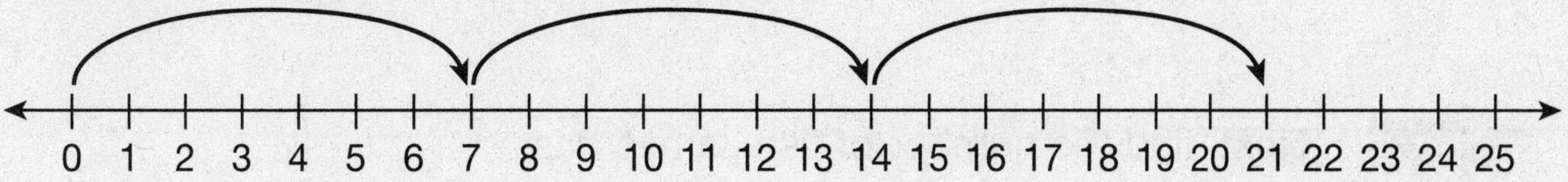

So, $3 \times 7 = 21$.

Solution **There are 21 people on the trail in all.**

Another phrase you might see in a word problem is *times as much*. This is another clue that you need to multiply.

Example 4

Sally's ribbon is 4 inches long. Tania's ribbon is 6 times as long as Sally's. How long is Tania's ribbon?

Strategy **Write a multiplication sentence. Then use repeated addition.**

Step 1 Write a multiplication sentence.

You know Sally's ribbon is 4 inches long and Tania's ribbon is 6 times as long.

6 times as long as 4 is the same as 6×4.

$6 \times 4 = \square$

Step 2 Add 6 four times.

$6 + 6 + 6 + 6 = 24$

So, $6 \times 4 = 24$.

Solution **Tania's ribbon is 24 inches long.**

Example 5

Daniel has 2 fish tanks. He has 12 fish in each tank. James has fewer fish than Daniel. Together they have 32 fish. How many fish does James have?

Strategy **Write number sentences to model the problem.**

Step 1 First, find the number of fish Daniel has.

Write and solve a multiplication sentence.
Use the symbol □ for the product.

You know Daniel has 2 fish tanks with 12 fish in each tank.

$2 \times 12 = \square$

$2 \times 12 = 24$

Step 2 Next, find the number of fish James has.

Write and solve a subtraction sentence.
Use the symbol □ for the difference.

You know that together they have 32 fish.

$32 - 24 = \square$

$32 - 24 = 8$

Solution **James has 8 fish.**

Coached Example

At Buddy's Bakery a cookie costs $2. A cake costs 4 times as much as a cookie. How much does a cake cost at Buddy's Bakery?

Write a multiplication sentence.

A cookie costs $________ and a cake costs ________ times as much.

________ × ________ = □

Use doubling to solve. Double ________.

________ + ________ = ________

So, ________ × ________ = ________.

A cake costs $________ at Buddy's Bakery.

Lesson Practice

Choose the correct answer.

1. Kelly baked 5 trays of muffins. Each tray holds 6 muffins. How many muffins did Kelly bake in all?

- ○ **A.** 11
- ○ **B.** 25
- ○ **C.** 30
- ○ **D.** 50

2. Mr. Field's garden has 8 rows of plants. Each row has 10 plants. How many plants does Mr. Field's garden have in all?

- ○ **A.** 18
- ○ **B.** 40
- ○ **C.** 70
- ○ **D.** 80

3. Steven bought 3 bags of potatoes. Each bag has 7 potatoes. How many potatoes did Steven buy in all?

- ○ **A.** 10
- ○ **B.** 21
- ○ **C.** 28
- ○ **D.** 30

4. Ebony has 7 bookshelves. She has 9 books on each shelf. Whitney has 12 more books than Ebony. How many books does Whitney have?

- ○ **A.** 75
- ○ **B.** 63
- ○ **C.** 51
- ○ **D.** 28

5. A toy car costs $5. A toy helicopter costs 3 times as much. How much does a toy helicopter cost?

- ○ **A.** $10
- ○ **B.** $15
- ○ **C.** $20
- ○ **D.** $30

6. Jesse's flower is 7 inches tall. Ted's flower is 2 times as tall as Jesse's. How tall is Ted's flower?

- ○ **A.** 7 inches
- ○ **B.** 9 inches
- ○ **C.** 10 inches
- ○ **D.** 14 inches

7. There are 5 parents driving the students from Ms. Alvarez's class to a play. There are 4 students in each car. How many students from Ms. Alvarez's class are going to the play?

- ○ **A.** 20
- ○ **B.** 24
- ○ **C.** 25
- ○ **D.** 30

8. There are 10 players on each basketball court. How many players are there on 6 basketball courts?

- ○ **A.** 30
- ○ **B.** 50
- ○ **C.** 60
- ○ **D.** 80

9. There are 4 lemon trees in Rasheed's backyard. There are 12 lemons growing on each tree.

A. Draw a model of the problem.

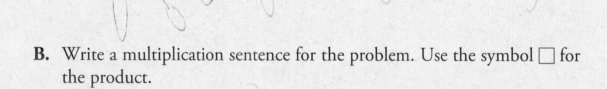

B. Write a multiplication sentence for the problem. Use the symbol □ for the product.

__________ × __________ = □

C. How many lemons in all are growing in Rasheed's backyard?

____

Common Core State Standard:
3.OA.5

Multiplication Properties

Getting the Idea

You can use multiplication properties to help you learn basic facts.

The **commutative property of multiplication** says that changing the order of factors does not change the product.

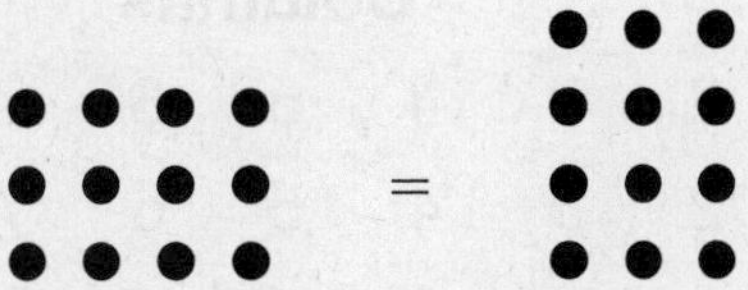

$3 \times 4 = 12$ $4 \times 3 = 12$

Example 1

What number makes the sentence true?

$2 \times 5 = \square \times 2$

Strategy **Use the commutative property of multiplication.**

The commutative property of multiplication says that changing the order of factors does not change the product.

$2 \times 5 = 10$, so $5 \times 2 = 10$

Solution **The number 5 makes the sentence true.**

You can use the commutative property to learn multiplication facts.

Look at the shaded row and column in the multiplication table below.

The multiplication facts for 3s have the same factors and products.

This is true for all multiplication facts.

For example, when you know the multiplication fact $3 \times 6 = 18$, you also know the multiplication fact $6 \times 3 = 18$.

Columns

Rows

×	1	2	3	4	5	6	7	8	9	10
1	1	2	3	4	5	6	7	8	9	10
2	2	4	6	8	10	12	14	16	18	20
3	3	6	9	12	15	18	21	24	27	30
4	4	8	12	16	20	24	28	32	36	40
5	5	10	15	20	25	30	35	40	45	50
6	6	12	18	24	30	36	42	48	54	60
7	7	14	21	28	35	42	49	56	63	70
8	8	16	24	32	40	48	56	64	72	80
9	9	18	27	36	45	54	63	72	81	90
10	10	20	30	40	50	60	70	80	90	100

The **distributive property of multiplication** says that multiplying a sum by a factor is the same as multiplying each addend by the factor and adding the products.

For example, use the distributive property to find 4×8.

Rename one of the factors.	$4 \times (5 + 3)$
Multiply the other factor by each addend.	$(4 \times 5) + (4 \times 3)$
Add the products.	$20 + 12 = 32$

Example 2

Find the product.

$6 \times 9 = \square$

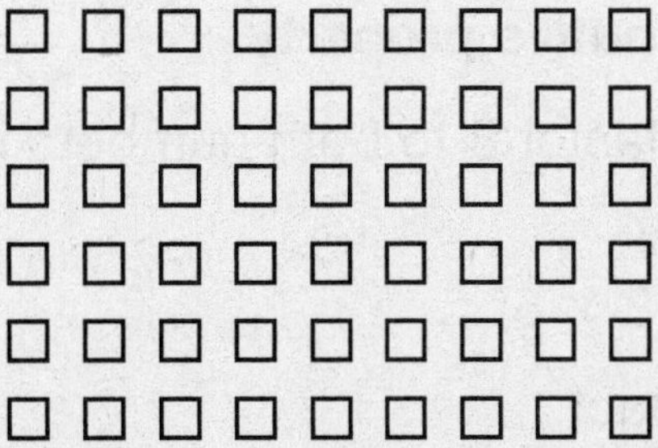

Strategy **Use the distributive property of multiplication.**

Step 1 Rename one of the factors.

Distribute the factor 6 to both numbers.

6×9

$6 \times (5 + 4)$

$(6 \times 5) + (6 \times 4)$

Step 2 Multiply each fact.

$(6 \times 5) + (6 \times 4)$

$30 + 24$

$6 \times 5 = 30$ $6 \times 4 = 24$

Step 3 Add the products.

$30 + 24 = 54$

Solution $\mathbf{6 \times 9 = 54}$

Example 3

Karen can't remember the product for 8 × 9, so she breaks the factor 9 into 4 + 5. What is 8 × (4 + 5)?

Strategy **Use the distributive property.**

Step 1 Distribute the factor 8 to both numbers inside the parentheses.

$8 \times (4 + 5)$

$(8 \times 4) + (8 \times 5)$

Step 2 Multiply each fact.

$(8 \times 4) + (8 \times 5)$

$32 + 40$

Step 3 Add the products.

$32 + 40 = 72$

Solution $\mathbf{8 \times 9 = 8 \times (4 + 5) = 72}$

Coached Example

4 × 9 = 36. What is the product of 9 × 4?

Look at 4 × 9 = 36.

The factors are ___________ and ___________.

Look at 9 × 4 = ☐.

The factors are ___________ and ___________.

The factors are the same, but the order of the factors is different.

The commutative property of multiplication says that changing the ___________________ of the factors does not change the product.

So, 9 × 4 = ___________.

The product of 9 × 4 is ___________.

Lesson Practice

Choose the correct answer.

1. Which is the missing number?

$4 \times 5 = \square \times 4$

- ○ **A.** 0
- ○ **B.** 1
- ○ **C.** 4
- ○ **D.** 5

2. Find the product.

$2 \times (4 + 6) = \square$

- ○ **A.** 20
- ○ **B.** 12
- ○ **C.** 10
- ○ **D.** 8

3. If $7 \times 4 = 28$, what is 4×7?

- ○ **A.** 21
- ○ **B.** 24
- ○ **C.** 27
- ○ **D.** 28

4. What number belongs in the $\square$?

$3 \times 8 = \square \times 3$

- ○ **A.** 3
- ○ **B.** 5
- ○ **C.** 8
- ○ **D.** 11

5. Which is the missing number?

$6 \times 9 = (6 \times 4) + (6 \times \square)$

- ○ **A.** 4
- ○ **B.** 5
- ○ **C.** 6
- ○ **D.** 9

6. What number belongs in the $\square$?

$1 \times \square = 1 \times 5$

- ○ **A.** 0
- ○ **B.** 1
- ○ **C.** 5
- ○ **D.** 25

7. Which has the same product as this sentence?

$7 \times (2 + 5) = \square$

○ A. $(7 + 2) + (7 + 5) = \square$

○ B. $(7 + 2) \times (7 \times 5) = \square$

○ C. $(7 \times 2) \times (7 \times 5) = \square$

○ D. $(7 \times 2) + (7 \times 5) = \square$

8. Which is the missing number?

$\square \times 6 = 6 \times 1$

○ A. 6

○ B. 2

○ C. 1

○ D. 0

9. Bianca cannot remember the product of 8×6. She decides to use the distributive property of multiplication to help.

A. Show how Bianca could rewrite the multiplication sentence using the distributive property.

5×6 and 3×6
____

B. Solve the problem. Show your work.

Common Core State Standard:
3.NBT.3

Multiply by Multiples of 10

Getting the Idea

A **multiple** of 10 is the product of 10 and any number.

All numbers that end with a 0 are multiples of 10, such as 10, 40, and 800.

You can use models to help you multiply multiples of 10.

Example 1

Find the product.

$6 \times 20 = \square$

Strategy **Use models.**

Step 1 Show 6 groups of 2 tens.

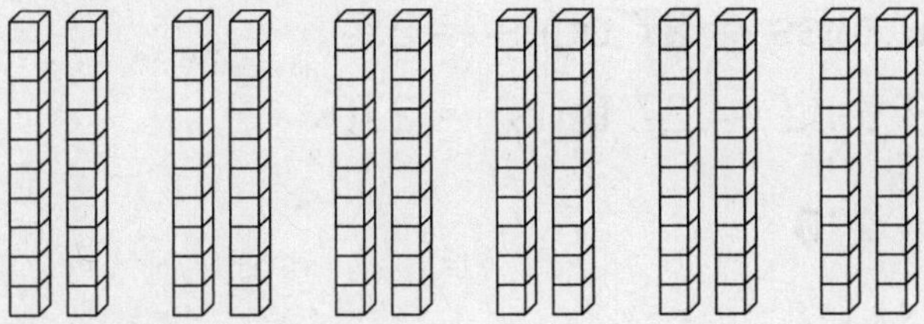

Step 2 Count the tens.

There are 12 tens.

12 tens = 120

Solution $\mathbf{6 \times 20 = 120}$

You can also use a basic fact and place value to find 6×20.

Think about the basic fact $6 \times 2 = 12$.

6×2 ones $= 12$ ones $= 12$

6×2 tens $= 12$ tens $= 120$

Example 2

Multiply.

$3 \times 90 = \square$

Strategy **Use a multiplication fact and place value.**

Step 1 Think of a basic fact.

$3 \times 9 = 27$

Step 2 Use place value.

3×9 ones $= 27$ ones $= 27$

3×9 tens $= 27$ tens $= 270$

Solution $\mathbf{3 \times 90 = 270}$

Example 3

Each marching band has 40 members.
How many members are in 5 marching bands?

Strategy **Use a basic fact and mental math.**

Step 1 Decide how to solve the problem.

Find 5 groups of 40.

So, find 5×40.

Step 2 Use a basic fact and place value.

Think: $5 \times 4 = 20$

5×4 ones $= 20$ ones $= 20$

5×4 tens $= 20$ tens $= 200$

Solution **There are 200 members in 5 marching bands.**

Coached Example

Rachel made 30 bags of treats. She put 5 treats in each bag. How many treats did Rachel bag in all?

Decide how to solve the problem.

Find 30 groups of _____.

So, find _____ $\times$ _____.

Use a basic fact.

Think: $3 \times 5 =$ _____

Use place value.

_____ ones $\times 5 =$ _____ ones $=$ _____

_____ tens $\times 5 =$ _____ tens $=$ _____

Rachel bagged _____ treats in all.

Lesson Practice

Choose the correct answer.

1. Find the product.

$2 \times 50 = \square$

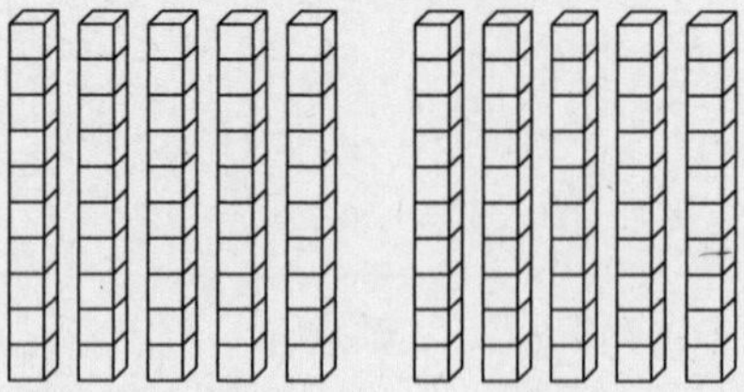

- ○ **A.** 52
- ○ **B.** 90
- ○ **C.** 100
- ○ **D.** 120

2. Find the product.

$3 \times 10 = \square$

- ○ **A.** 3
- ○ **B.** 10
- ○ **C.** 30
- ○ **D.** 300

3. Find the product.

$7 \times 50 = \square$

- ○ **A.** 35
- ○ **B.** 75
- ○ **C.** 120
- ○ **D.** 350

4. Find the product.

$80 \times 4 = \square$

- ○ **A.** 320
- ○ **B.** 240
- ○ **C.** 120
- ○ **D.** 32

5. Find the product.

$70 \times 8 = \square$

- ○ **A.** 56
- ○ **B.** 78
- ○ **C.** 150
- ○ **D.** 560

6. Ali delivers 30 newspapers each day. How many papers does she deliver in 6 days?

- ○ **A.** 150
- ○ **B.** 180
- ○ **C.** 200
- ○ **D.** 210

7. Mr. Davis drives 20 miles round-trip for work each day. How many miles does he drive for work in 5 days?

- ○ **A.** 10 miles
- ○ **B.** 70 miles
- ○ **C.** 100 miles
- ○ **D.** 170 miles

8. Which sentence does **not** have the same product as the others?

- ○ **A.** 4 × 90 = ☐
- ○ **B.** 5 × 70 = ☐
- ○ **C.** 6 × 60 = ☐
- ○ **D.** 9 × 40 = ☐

9. Maureen can type 40 words a minute. She wants to know how many words she can type in 9 minutes.

A. Which basic fact can you use to find how many words Maureen can type in 9 minutes?

____

B. How many words can Maureen type in 9 minutes? Show your work.

____

____

Common Core State Standard:
3.OA.5

Multiply Three Numbers

Getting the Idea

Sometimes you may have to multiply three factors to find a product. When this happens, first multiply two of the factors, and then multiply the product of those two factors by the third factor.

For example, find the product of $6 \times 3 \times 4$.

First multiply two factors.

$6 \times 3 = 18$

Then multiply the product of the two factors by the other factor.

$18 \times 4 = 72$

So, $6 \times 3 \times 4 = 72$.

You can use models to find the product of 3 factors.

Example 1

Find the product.

$2 \times 3 \times 5 = \square$

Strategy **Use models.**

Step 1 Multiply two factors.

2 rows of 3 counters

$2 \times 3 = 6$

Step 2 Multiply the product by the other factor.

2 rows of 3 counters 5 times

$6 \times 5 = 30$

Solution $\mathbf{2 \times 3 \times 5 = 30}$

Another way to multiply $2 \times 3 \times 5$ is to use the commutative property. Remember, the commutative property says that you can multiply factors in any order, and the product will not change.

Multiply $2 \times 3 \times 5$.

Change the order of factors 3 and 5. $2 \times \mathbf{5} \times \mathbf{3}$

Multiply. $10 \times 3 = 30$

So, $2 \times 3 \times 5 = 30$.

You can also group 3 factors in different ways to help you multiply.

The **associative property of multiplication** says that changing the grouping of the factors does not change the product.

For example, $(2 \times 4) \times 3 = 2 \times (4 \times 3)$.

$(2 \times 4) \times 3 = 8 \times 3$ $\quad$ $2 \times (4 \times 3) = 2 \times 12$

$8 \times 3 = 24$ $\quad$ $2 \times 12 = 24$

Example 2

Find the product.

$(7 \times 5) \times 2 = \square$

Strategy **Use the associative property of multiplication.**

Step 1 Change the grouping of the factors.

Use mental math. Think: $5 \times 2 = 10$

$(7 \times 5) \times 2 = 7 \times (5 \times 2)$

Step 2 Multiply inside the parentheses.

$7 \times (5 \times 2)$

7×10

Step 3 Multiply the product by the other factor.

$7 \times 10 = 70$

Solution $\mathbf{(7 \times 5) \times 2 = 70}$

Coached Example

Find the product.

(6 × 2) × 3 = □

Use the associative property of multiplication.

Change the grouping of the factors to help you multiply.

(6 × 2) × 3 = ________ × (________ × ________)

Multiply inside the parentheses.

(________ × ________) = ________

Multiply the product by the other factor.

________ × ________ = ________

(6 × 2) × 3 = ________

Lesson Practice

Choose the correct answer.

1. Which shows the associative property of multiplication?

- ○ **A.** $(2 \times 3) \times 4 = 2 \times (3 \times 4)$
- ○ **B.** $(5 \times 9) = (9 \times 5)$
- ○ **C.** $4 \times 1 = 4$
- ○ **D.** $(3 \times 4) = (2 \times 6)$

2. Find the product.

$4 \times 5 \times 3 = \square$

- ○ **A.** 4
- ○ **B.** 8
- ○ **C.** 12
- ○ **D.** 60

3. Find the product.

$2 \times (4 \times 2) = \square$

- ○ **A.** 6
- ○ **B.** 8
- ○ **C.** 10
- ○ **D.** 16

4. Find the product.

$(3 \times 4) \times 3 = \square$

- ○ **A.** 12
- ○ **B.** 24
- ○ **C.** 36
- ○ **D.** 48

5. Which shows another way to group $(6 \times 4) \times 2$?

- ○ **A.** $(6 \times 4) + (6 \times 2)$
- ○ **B.** $6 \times (4 \times 2)$
- ○ **C.** $(12 \times 2) \times 4$
- ○ **D.** $(6 + 4) \times (4 + 2)$

6. Find the product.

$8 \times (2 \times 3) = \square$

- ○ **A.** 6
- ○ **B.** 16
- ○ **C.** 24
- ○ **D.** 48

7. Find the product.

$5 \times 2 \times 6 = \square$

- A. 60
- B. 30
- C. 12
- D. 10

8. Find the product.

$(9 \times 1) \times 8 = \square$

- A. 8
- B. 9
- C. 18
- D. 72

9. Rachael wants to solve $(4 \times 3) \times 2$. Her teacher told her to change the grouping of the factors to help her multiply.

A. Write another way that Rachael can group the factors.

____

B. Find the product of $(4 \times 3) \times 2$. Show your work.

____

Common Core State Standards: 3.OA.2, 3.OA.3, 3.OA.4, 3.OA.7

Understand Division

Getting the Idea

You can use **division** to find the number of equal groups or the number in each equal group.

Here are the parts of a division sentence.

6	÷	3	=	2
dividend		**divisor**		**quotient**

Example 1

Find the quotient.

18 ÷ 2 = □

Strategy **Draw a picture.**

Step 1 Make 18 circles to show 18.

Make 2 equal groups.

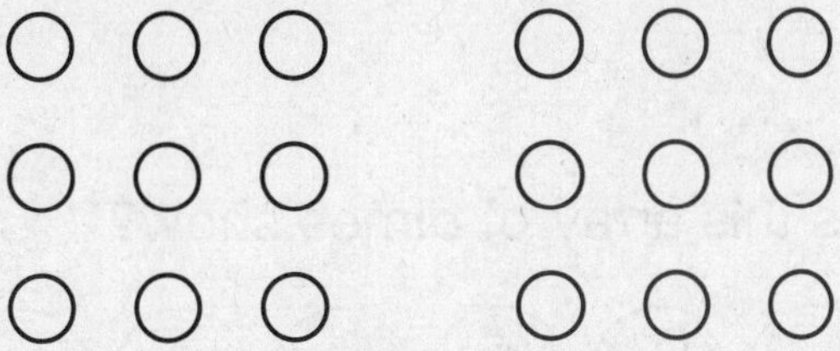

Step 2 Count how many are in each group.

There are 9 circles in each group.

Solution **18 ÷ 2 = 9**

You can use **repeated subtraction** to find the quotient.

Example 2

Find the quotient.

$12 \div 3 = \square$

Strategy **Use repeated subtraction.**

Step 1 Start with 12. Subtract 3 until you reach 0.

$12 - 3 = 9$

$9 - 3 = 6$

$6 - 3 = 3$

$3 - 3 = 0$

Step 2 Count the number of times you subtracted 3.

You subtracted 4 times.

Solution $\mathbf{12 \div 3 = 4}$

You can use an array to find the number of equal groups.

Example 3

What division facts does this array of dimes show?

Strategy **Count the number of dimes, rows, and dimes in each row.**

Step 1 Count the total number of dimes.

There are 32 dimes.

Step 2 Count the number of rows.

There are 4 rows.

Step 3 Count the number of dimes in each row.

There are 8 dimes in each row.

Step 4 Write the division facts.

32	÷	4	=	8
total number of dimes		number of rows		number in each row

32	÷	8	=	4
total number of dimes		number in each row		number of rows

Solution **The array of dimes shows 32 ÷ 4 = 8 and 32 ÷ 8 = 4.**

Example 4

What division facts does this area model show?

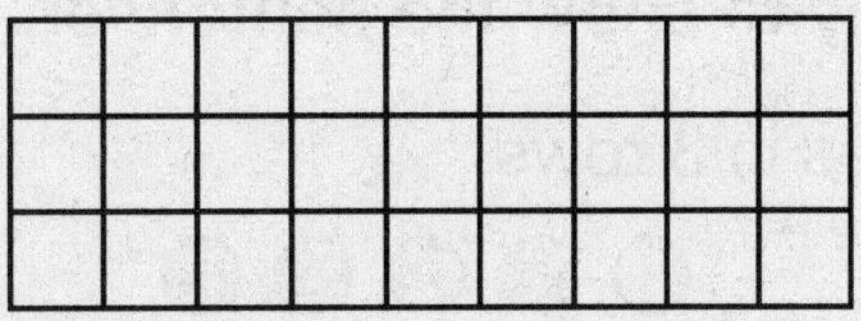

Strategy **Count the number of squares, rows, and squares in each row.**

Step 1 Count the total number of squares.

There are 27 squares in all.

Step 2 Count the number of rows.

There are 3 rows of squares.

Step 3 Count the number of squares in each row.

There are 9 squares in each row.

Step 4 Write the division facts.

27 ÷ 3 = 9 and 27 ÷ 9 = 3

Solution **The area model shows 27 ÷ 3 = 9 and 27 ÷ 9 = 3.**

Multiplication and division are **inverse operations**, or opposites.

Inverse operations undo each other. So you can use a multiplication fact to solve a division fact, or a division fact to solve a multiplication fact.

A **fact family** is a group of related facts that use the same numbers.

Here is the fact family for 2, 3, and 6.

$3 \times 2 = 6$	$2 \times 3 = 6$
$6 \div 3 = 2$	$6 \div 2 = 3$

Example 5

These two sentences are in the same fact family.

$3 \times \square = 15$

$15 \div \square = 3$

What number makes both sentences true?

Strategy **Make an array to show the sentences.**

Step 1 Draw 15 circles in 3 rows.

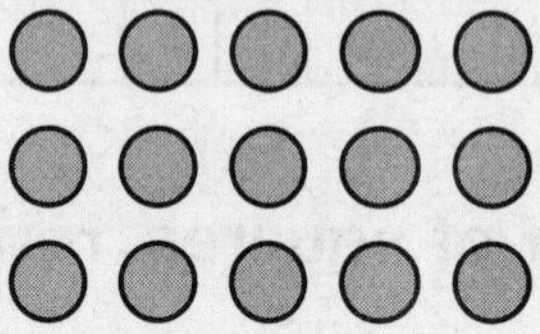

Step 2 Find the missing number in $3 \times \square = 15$.

3 rows of 5 counters equal 15.

$3 \times 5 = 15$

Step 3 Find the missing number in $15 \div \square = 3$.

The 15 counters are in 3 equal rows of 5.

$15 \div 5 = 3$

Solution **The number 5 makes both sentences true.**

$\mathbf{3 \times 5 = 15}$

$\mathbf{15 \div 5 = 3}$

Coached Example

What multiplication-division fact family does this picture show?

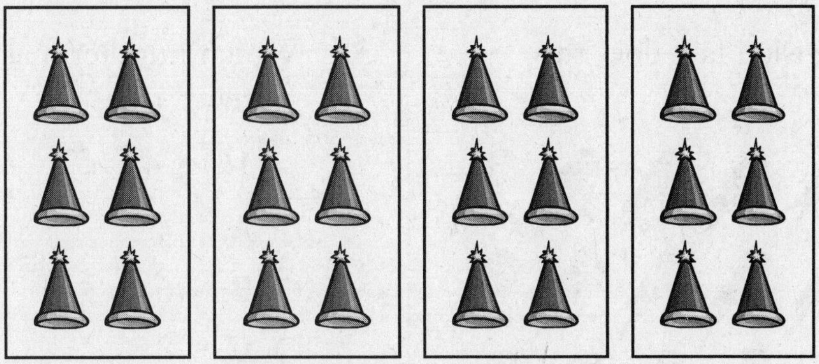

How many equal groups of hats are there? __________

How many hats are in each group? __________

How many hats are there in all? __________

Write the multiplication facts for this picture.

4 × __________ = __________

__________ × __________ = __________

Write the division facts for this picture.

24 ÷ __________ = __________

__________ ÷ __________ = __________

Lesson Practice

Choose the correct answer.

1. Which division fact does this array show?

- ○ **A.** $15 \div 3 = 5$
- ○ **B.** $20 \div 5 = 4$
- ○ **C.** $21 \div 3 = 7$
- ○ **D.** $25 \div 5 = 5$

2. Which division fact does this picture show?

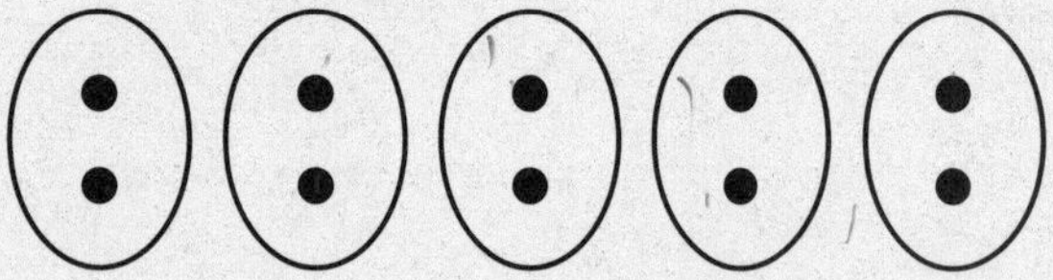

- ○ **A.** $20 \div 2 = 10$
- ○ **B.** $10 \div 5 = 2$
- ○ **C.** $5 \div 5 = 1$
- ○ **D.** $5 \div 1 = 5$

3. Which number makes this sentence true?

$16 \div 4 = \square$

- ○ **A.** 4
- ○ **B.** 6
- ○ **C.** 8
- ○ **D.** 12

4. Which fact is **not** related to any of the other facts?

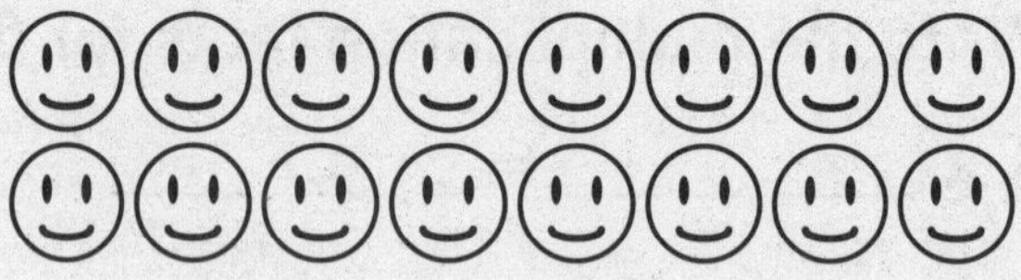

- ○ **A.** $16 \div 2 = 8$
- ○ **B.** $2 \times 8 = 16$
- ○ **C.** $8 \times 2 = 16$
- ○ **D.** $16 \div 4 = 4$

5. Which number makes both sentences true?

$40 \div \square = 4$

$4 \times \square = 40$

- ○ **A.** 10
- ○ **B.** 9
- ○ **C.** 8
- ○ **D.** 7

6. Which multiplication fact can be used to find the missing number?

$36 \div \square = 9$

- ○ **A.** $2 \times 18 = 36$
- ○ **B.** $6 \times 6 = 36$
- ○ **C.** $9 \times 4 = 36$
- ○ **D.** $36 \times 1 = 36$

7. Which division fact does this area model show?

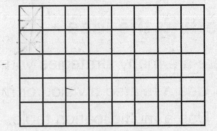

- ○ **A.** $60 \div 5 = 12$
- ○ **B.** $50 \div 10 = 5$
- ○ **C.** $40 \div 4 = 10$
- ○ **D.** $40 \div 5 = 8$

8. Find the quotient.

$48 \div 8 = \square$

- ○ **A.** 6
- ○ **B.** 7
- ○ **C.** 8
- ○ **D.** 9

9. Dennis drew the picture below.

A. Write two division facts for the picture.

____

____

B. Write two multiplication facts for the picture.

____

____

Division Facts

Common Core State Standards: 3.OA.3, 3.OA.4, 3.OA.6, 3.OA.7

Getting the Idea

There are many strategies you can use to solve division problems.

- Use a related division or multiplication fact.
- Use a multiplication table.
- Skip count backward.
- Use repeated subtraction.
- Make a model.

Example 1

Find the quotient.

$40 \div 8 = \square$

Strategy **Use a related multiplication fact.**

Step 1 Look at the numbers in the division problem.

40 and 8

Step 2 Use a related multiplication fact.

Think: $8 \times ? = 40$

$8 \times 5 = 40$

Step 3 Multiplication and division are inverse operations.

So, $40 \div 8 = 5$.

Solution $\mathbf{40 \div 8 = 5}$

You can use a multiplication table to help you with basic division facts.

Example 2

Find the quotient.

36 ÷ 9 = □

Strategy **Use a multiplication table.**

Step 1 Look at the 9s row.

Find 36.

Step 2 From 36, go to the top to find which column it is in.

It is in the 4s column.

The quotient is 4.

×	**0**	**1**	**2**	**3**	**4**	**5**	**6**	**7**	**8**	**9**	**10**
0	0	0	0	0	0	0	0	0	0	0	0
1	0	1	2	3	4	5	6	7	8	9	10
2	0	2	4	6	8	10	12	14	16	18	20
3	0	3	6	9	12	15	18	21	24	27	30
4	0	4	8	12	16	20	24	28	32	36	40
5	0	5	10	15	20	25	30	35	40	45	50
6	0	6	12	18	24	30	36	42	48	54	60
7	0	7	14	21	28	35	42	49	56	63	70
8	0	8	16	24	32	40	48	56	64	72	80
9	0	9	18	27	**36**	45	54	63	72	81	90
10	0	10	20	30	40	50	60	70	80	90	100

Solution **36 ÷ 9 = 4**

You can also use skip counting to solve division problems. Start from the total and skip count backward until you reach 0.

Example 3

Find the quotient.

15 ÷ 5 = □

Strategy **Skip count backward.**

Step 1 Skip count backward by 5s from 15 to 0.

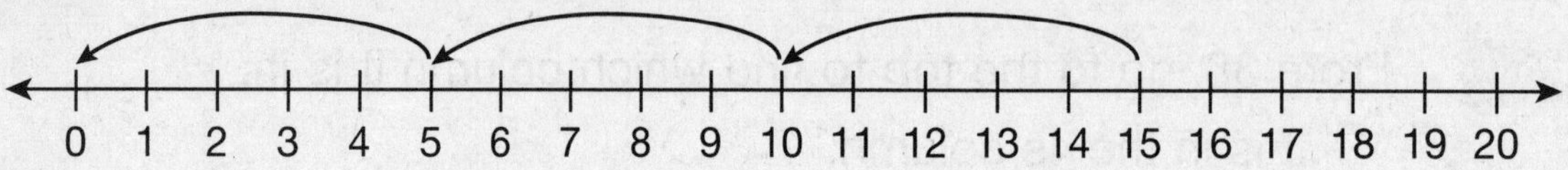

Step 2 Count how many times you skip counted backward.

You skip counted 3 times.

Solution **15 ÷ 5 = 3**

You can use repeated subtraction to solve division problems.

Example 4

Find the quotient.

27 ÷ 9 = □

Strategy **Use repeated subtraction.**

Step 1 Start with 27. Subtract 9 until you reach 0.

27 − 9 = 18

18 − 9 = 9

9 − 9 = 0

Step 2 Count how many times you subtracted the number 9.

The number 9 was subtracted 3 times.

Solution **27 ÷ 9 = 3**

Coached Example

Write the division facts for this array.

Count the rows, the number of dots in each row, and the total number of dots.

How many rows? ____________

How many dots in each row? ____________

How many dots in all? ____________

Write the division facts.

____________ ÷ ____________ = ____________

____________ ÷ ____________ = ____________

Lesson Practice

Choose the correct answer.

1. Find the quotient.

$32 \div 4 = \square$

- ○ **A.** 16
- ○ **B.** 12
- ○ **C.** 9
- ○ **D.** 8

2. Find the quotient.

$70 \div 7 = \square$

- ○ **A.** 7
- ○ **B.** 10
- ○ **C.** 63
- ○ **D.** 77

3. What is $81 \div 9$?

- ○ **A.** 6
- ○ **B.** 7
- ○ **C.** 8
- ○ **D.** 9

4. Find the quotient.

$28 \div 7 = \square$

- ○ **A.** 11
- ○ **B.** 6
- ○ **C.** 4
- ○ **D.** 1

5. Find the quotient.

$90 \div 10 = \square$

- ○ **A.** 8
- ○ **B.** 9
- ○ **C.** 10
- ○ **D.** 11

6. Find the quotient.

$42 \div 7 = \square$

- ○ **A.** 6
- ○ **B.** 8
- ○ **C.** 9
- ○ **D.** 49

7. What is 45 ÷ 5?

○ A. 8

○ B. 9

○ C. 10

○ D. 40

8. Find the quotient.

$24 \div 2 = \square$

○ A. 26

○ B. 22

○ C. 12

○ D. 10

9. Mark wants to find the quotient for the following division fact.

$32 \div 8 = \square$

A. Use repeated subtraction to find the quotient. Show your work.

____

____

____

____

____

B. Write two related multiplication facts for this division problem.

____

____

Division Word Problems

Common Core State Standards:
3.OA.3, 3.OA.8

Getting the Idea

With division word problems, look for key phrases to help you solve the problem.

How many in each group or *how many groups* means you need to find the divisor or the quotient.

How many in all means you need to find the dividend.

When you write a division sentence, remember to use a symbol or letter to represent the unknown number.

Example 1

Tony wants to share 12 pencils equally among 3 friends. How many pencils will each friend get?

Strategy **Write a division sentence for the problem. Then draw a picture to solve.**

Step 1 Write a division sentence. Use the symbol □ for the quotient.

There are 12 pencils. There are 3 friends.

total number of pencils	÷	number of friends	=	number of pencils each friend will get

$12 \div 3 = \square$

Step 2 Draw a picture.

Draw 12 pencils. Circle 3 equal groups.

There are 4 pencils in each group.

$12 \div 3 = 4$

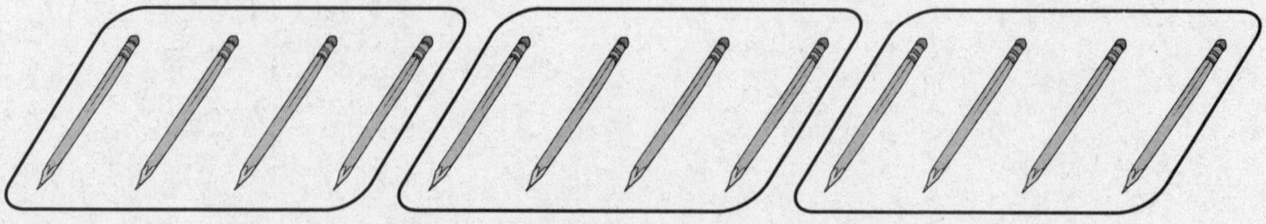

Solution **The three friends will get 4 pencils each.**

Example 2

Casey picked 20 apples. She gave 4 apples each to some friends. She does not have any apples left. How many friends received apples from Casey?

Strategy **Write a division sentence. Then use repeated subtraction to solve.**

Step 1 Write a division sentence. Use □ for the quotient.

There are 20 apples. Each friend got 4 apples.

total number of apples	÷	number of apples in each group	=	number of friends

$20 \div 4 = \square$

Step 2 Use repeated subtraction.

Start with 20. Subtract 4 each time until you reach 0.

$20 - 4 = 16$

$16 - 4 = 12$

$12 - 4 = 8$

$8 - 4 = 4$

$4 - 4 = 0$

Step 3 Count the number of times you subtracted 4.

You subtracted 5 times.

$20 \div 4 = 5$

Solution **Five friends received 4 apples each from Casey.**

Example 3

Mr. Frey has 24 students. He seated the students at 4 tables. Each table had the same number of students. How many students were at each table?

Strategy **Write a division sentence. Then use a related multiplication fact to solve.**

Step 1 Write a division sentence. Use ☐ for the quotient.

There are 24 students in all. There are 4 equal groups of students.

total number ÷ number of groups = number in each group

$24 \div 4 = \square$

Step 2 Use a related multiplication fact.

Think: $4 \times ? = 24$

$4 \times 6 = 24$

Step 3 Multiplication and division are inverse operations.

So, $24 \div 4 = 6$.

Solution **Six students were at each table.**

Another phrase you might see in a word problem is *times as much*. This is another clue that you may need to divide.

Coached Example

A dozen flowers costs $28 and a plant costs $7. How many times as much does a dozen flowers cost as a plant?

Write a division sentence.

You know a dozen flowers costs $________ and a plant costs $________.

__28__ ÷ __7__ = ☐ 4

Use a related multiplication fact to help you solve the problem.

Think: 7 × __4__ = 28

Multiplication and division are inverse operations.

So, 28 ÷ 7 = __4__.

A dozen flowers costs __4__ times as much as a plant.

Lesson Practice

Choose the correct answer.

1. Nick has 48 DVDs in his collection. He keeps 6 DVDs on each shelf in a cabinet. How many shelves does Nick use for his DVDs?

- ○ **A.** 42
- ○ **B.** 8
- ○ **C.** 7
- ○ **D.** 6

2. Three friends share 30 marbles. Each friend gets the same number of marbles. How many marbles does each friend get?

- ○ **A.** 3
- ○ **B.** 4
- ○ **C.** 10
- ○ **D.** 27

3. Mrs. Martinez gave her 5 children $25 to share equally. How much money did each child receive?

- ○ **A.** $4
- ○ **B.** $5
- ○ **C.** $6
- ○ **D.** $20

4. Emma had 18 extra comic books to share. She divided them equally among 3 friends. How many comic books did each friend get?

- ○ **A.** 3
- ○ **B.** 6
- ○ **C.** 9
- ○ **D.** 15

5. Brenna has 16 flowers. She puts the same number of flowers into 4 bouquets. How many flowers are in each bouquet?

- ○ **A.** 4
- ○ **B.** 12
- ○ **C.** 32
- ○ **D.** 64

6. A bag of apples costs $6 and that is 3 times as much as a box of blueberries. How much does a box of blueberries cost?

- ○ **A.** $18
- ○ **B.** $12
- ○ **C.** $3
- ○ **D.** $2

7. There are 32 students who signed up for a clean-up project. They formed teams of 8 students each. How many teams did they form?

- ○ **A.** 4
- ○ **B.** 6
- ○ **C.** 24
- ○ **D.** 40

8. Will's toy train is 9 inches long. Lane's toy train is 36 inches long. How many times longer is Lane's train than Will's train?

- ○ **A.** 27
- ○ **B.** 18
- ○ **C.** 9
- ○ **D.** 4

9. Lilly baked 40 cookies. She shared her cookies equally among 4 friends. How many cookies did each friend receive?

A. Draw a model of the problem.

B. Write a division sentence for the problem. Use □ for the quotient.

______ ÷ ______ = □

C. How many cookies did each friend receive?

____

Domain 2: Cumulative Assessment for Lessons 9–18

1. Which is equal to 8×4?

   ○ **A.** $4 + 4 + 4 + 4$

   ○ **B.** $8 + 8 + 8 + 8$

   ○ **C.** $8 + 8$

   ○ **D.** $8 + 4 + 8 + 4$

2. Which shows the associative property of multiplication?

   ○ **A.** $1 \times 8 = 8$

   ○ **B.** $(5 \times 6) = (3 \times 10)$

   ○ **C.** $(5 \times 6) = (6 \times 5)$

   ○ **D.** $3 \times (5 \times 4) = (3 \times 5) \times 4$

3. Find the quotient.

   $32 \div 4 = \square$

   ○ **A.** 28

   ○ **B.** 8

   ○ **C.** 6

   ○ **D.** 4

4. Which number makes both sentences true?

   $36 \div \square = 4$

   $4 \times \square = 36$

   ○ **A.** 10

   ○ **B.** 9

   ○ **C.** 8

   ○ **D.** 7

5. Tasha has 5 sheets of stickers. Each sheet has 12 stickers on it. How many stickers does Tasha have in all?

   ○ **A.** 60

   ○ **B.** 55

   ○ **C.** 50

   ○ **D.** 17

6. Which sentence is true?

   ○ **A.** A number times 4 could be odd or even.

   ○ **B.** A number times 6 is always an even number.

   ○ **C.** A number times 7 is always an even number.

   ○ **D.** A number times 8 could be odd or even.

7. Which division fact does this picture show?

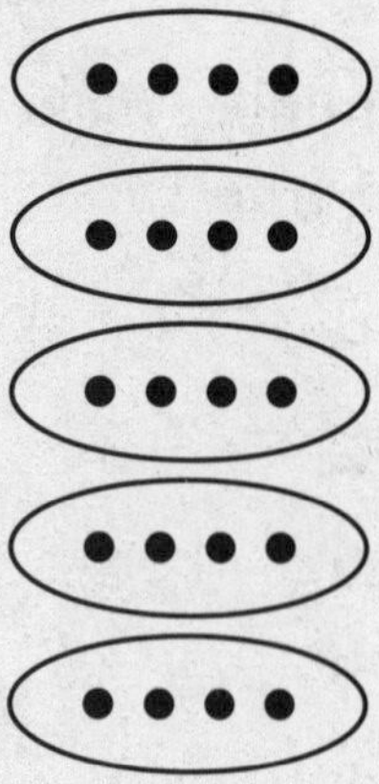

○ **A.** 20 ÷ 2 = 10

○ **B.** 20 ÷ 5 = 4

○ **C.** 15 ÷ 3 = 5

○ **D.** 20 ÷ 1 = 20

8. Which multiplication fact can be used to find the missing number?

42 ÷ □ = 7

○ **A.** 2 × 21 = 42

○ **B.** 3 × 14 = 42

○ **C.** 6 × 7 = 42

○ **D.** 42 × 1 = 42

9. Each Ferris wheel ride can seat 40 people. The Ferris wheel gives 6 rides each hour. What is the greatest number of people that can ride the Ferris wheel in an hour?

____

10. Mrs. Wagner has 8 bookshelves in her classroom. Each shelf has 7 books on it.

A. Draw a model of the problem.

B. Write a multiplication sentence for the problem. Use the symbol □ for the product.

__________ × __________ = □

C. How many books are on Mrs. Wagner's bookshelves?

____

Domain 3

Number and Operations—Fractions

Domain 3: Diagnostic Assessment for Lessons 19–22

Domain 3: Cumulative Assessment for Lessons 19–22

Domain 3: Diagnostic Assessment for Lessons 19–22

1. Where is point P located on the number line?

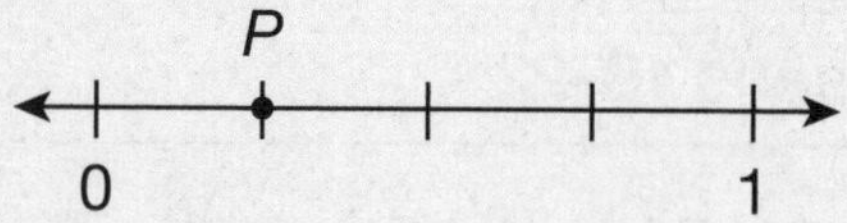

○ **A.** $\frac{1}{4}$

○ **B.** $\frac{2}{4}$

○ **C.** $\frac{3}{4}$

○ **D.** $\frac{4}{4}$

2. Which two fractions are equivalent?

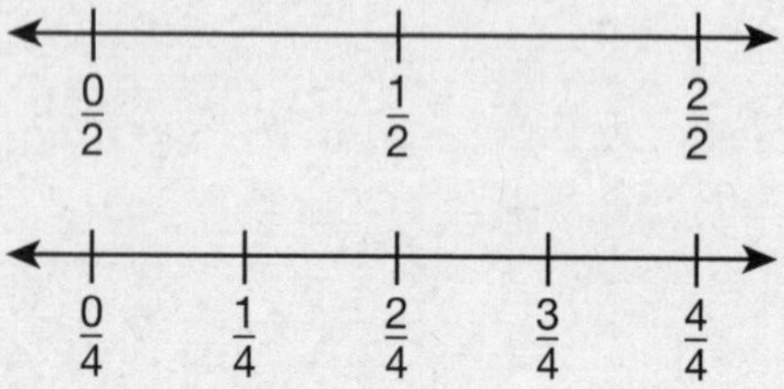

○ **A.** $\frac{1}{2}$ and $\frac{1}{4}$

○ **B.** $\frac{1}{2}$ and $\frac{2}{4}$

○ **C.** $\frac{2}{2}$ and $\frac{3}{4}$

○ **D.** $\frac{1}{2}$ and $\frac{4}{4}$

3. What fraction of the rectangle is shaded?

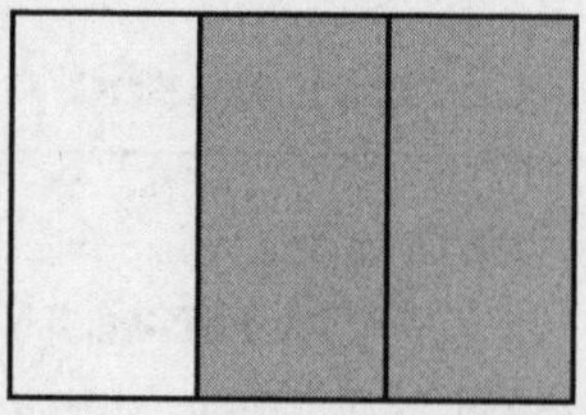

○ **A.** $\frac{1}{4}$

○ **B.** $\frac{1}{3}$

○ **C.** $\frac{2}{4}$

○ **D.** $\frac{2}{3}$

4. Which is the same as $\frac{4}{4}$?

○ **A.** 4

○ **B.** $\frac{4}{1}$

○ **C.** 1

○ **D.** $\frac{1}{4}$

5. Which symbol belongs in the ◯ to make the sentence true?

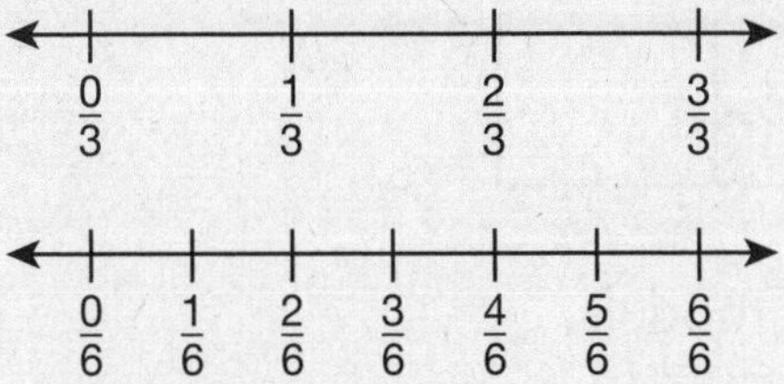

$\frac{3}{3} \bigcirc \frac{6}{6}$

○ **A.** >

○ **B.** <

○ **C.** =

○ **D.** +

6. Which fraction is equivalent to $\frac{2}{6}$?

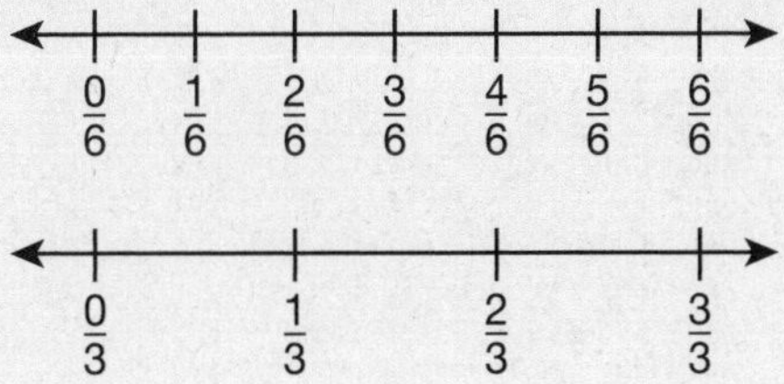

○ **A.** $\frac{2}{3}$

○ **B.** $\frac{1}{3}$

○ **C.** $\frac{1}{4}$

○ **D.** $\frac{1}{6}$

7. Which symbol belongs in the ◯ to make the sentence true?

$\frac{4}{8} \bigcirc \frac{6}{8}$

○ **A.** <

○ **B.** >

○ **C.** =

○ **D.** +

8. Which is another way to represent the number 3?

○ **A.** $\frac{3}{1}$

○ **B.** $\frac{1}{1}$

○ **C.** $\frac{3}{3}$

○ **D.** $\frac{1}{3}$

9. Draw point *B* at $\frac{4}{8}$ on the number line below.

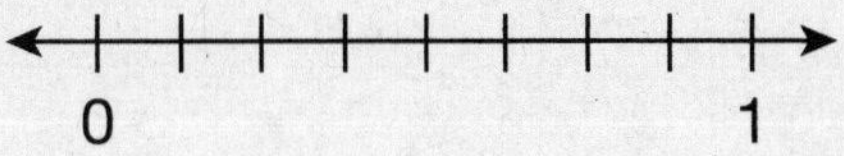

10. Gaia drew the models below.

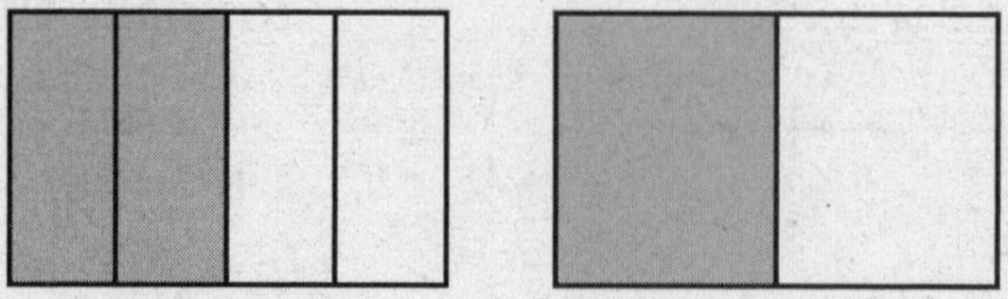

A. Write two equivalent fractions for Gaia's models.

_______________ and _______________

B. Show the two equivalent fractions on the number lines below.

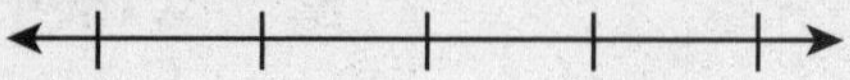

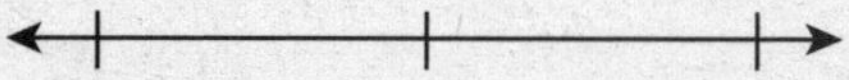

Fractions

Getting the Idea

A **fraction** names part of a whole. The **numerator** is the top number in a fraction. It tells the number of equal parts included in the fraction.

The **denominator** is the bottom number in a fraction. It tells the total number of equal parts that make up the whole.

The circle shows 3 equal parts.

Each part is $\frac{1}{3}$ of the circle.

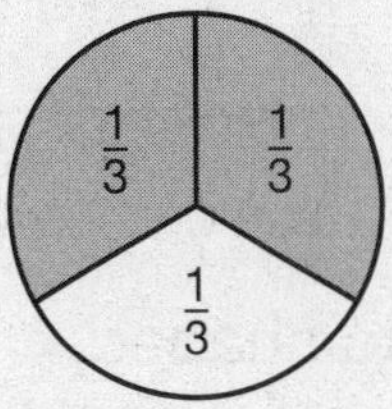

$\frac{\text{number of shaded parts}}{\text{total number of equal parts}}$ ⟵ $\frac{2}{3}$

$\frac{2}{3}$ of the circle is shaded.

The fraction $\frac{2}{3}$ is read as *two-thirds*.

Example 1

What part of the hexagon is shaded?

How is the fraction read?

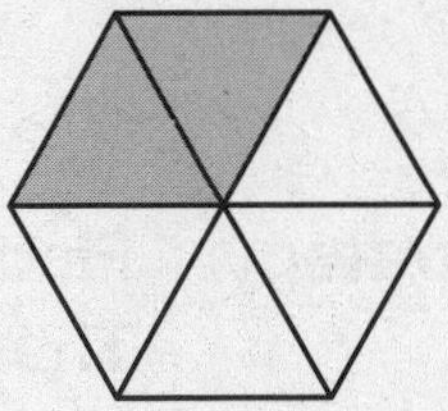

Strategy **Find the denominator and the numerator.**

Step 1 Count the total number of parts.

There are 6 parts. This is the denominator.

Step 2 Count the number of shaded parts.

There are 2 shaded parts. This is the numerator.

Step 3 Write the fraction.

$\frac{\text{numerator}}{\text{denominator}} = \frac{2}{6}$

Solution $\frac{2}{6}$ **of the hexagon is shaded. It is read as *two-sixths*.**

A fraction with 1 as its numerator is called a **unit fraction.**

All fractions are built from unit fractions, such as $\frac{1}{6}$.

For example, the number line below shows one whole from 0 to 1.

The whole is broken into 6 equal parts. Each equal part is $\frac{1}{6}$ of the whole.

There is a point located at $\frac{4}{6}$. So there are 4 parts of $\frac{1}{6}$ in $\frac{4}{6}$.

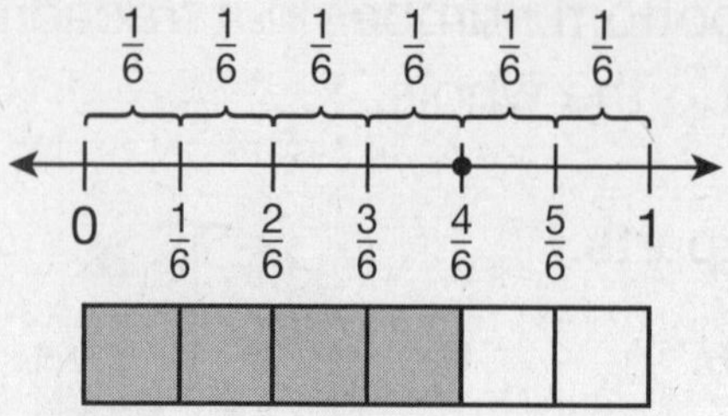

The rectangle shows $\frac{4}{6}$ parts shaded.

The fraction $\frac{4}{6}$ is built by combining 4 of the unit fraction $\frac{1}{6}$.

Example 2

What fraction is located at point *A* on the number line?

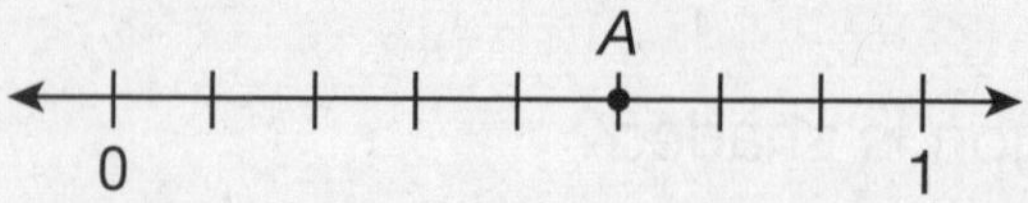

Strategy **Find how many equal parts the number line is divided into. Then find the denominator and the numerator.**

Step 1 Count the number of equal parts between 0 and 1.

There are 8 parts. This is the denominator.

Step 2 Count the number of parts between 0 and point *A*.

There are 5 parts. This is the numerator.

Step 3 Write the numerator over the denominator.

$$\frac{\text{numerator}}{\text{denominator}} = \frac{5}{8}$$

Solution **Point *A* is located at $\frac{5}{8}$ on the number line.**

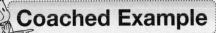

Coached Example

The figure below represents a sandwich that was served for lunch. The shaded part shows the amount of the sandwich that was eaten.

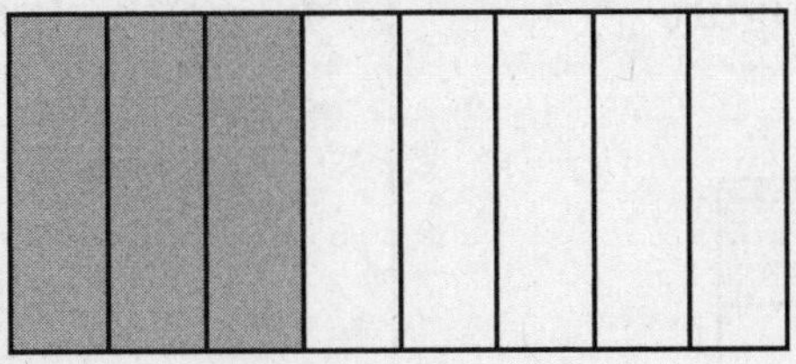

What fraction of the sandwich was eaten?

How many equal parts make up the figure? __________

This is the ____________________ of the fraction.

How many parts of the figure are shaded? __________

This is the ____________________ of the fraction.

Write the fraction.

____ ← numerator
____ ← denominator

So, __________ of the sandwich was eaten.

Lesson Practice

Choose the correct answer.

1. What fraction of the figure is shaded?

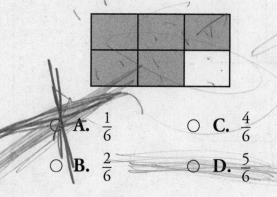

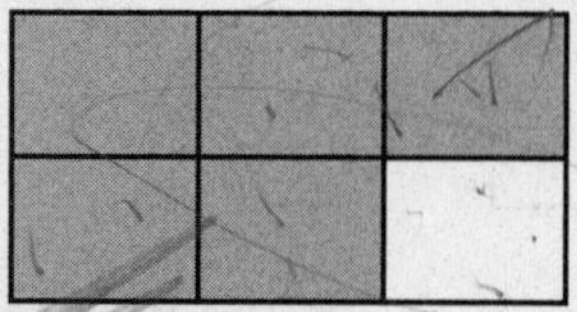

○ **A.** $\frac{1}{6}$ ○ **C.** $\frac{4}{6}$

○ **B.** $\frac{2}{6}$ ○ **D.** $\frac{5}{6}$

2. Which fraction has 5 for a numerator?

○ **A.** $\frac{1}{4}$ ○ **C.** $\frac{4}{7}$

○ **B.** $\frac{3}{5}$ ○ **D.** $\frac{5}{8}$

3. Where is point *B* located on the number line?

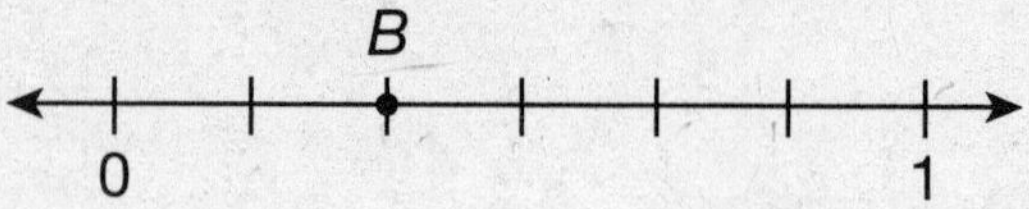

○ **A.** $\frac{1}{6}$ ○ **C.** $\frac{4}{6}$

○ **B.** $\frac{2}{6}$ ○ **D.** $\frac{5}{6}$

4. Which figure shows $\frac{3}{4}$ shaded?

○ **A.**

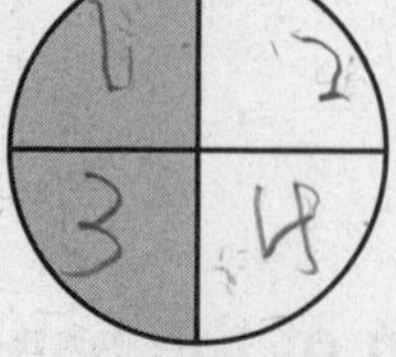

○ **B.**

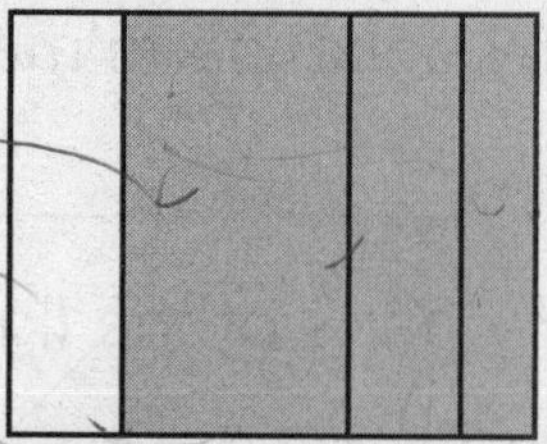

○ **C.**

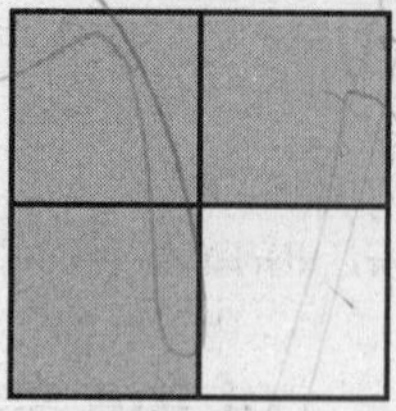

○ **D.**

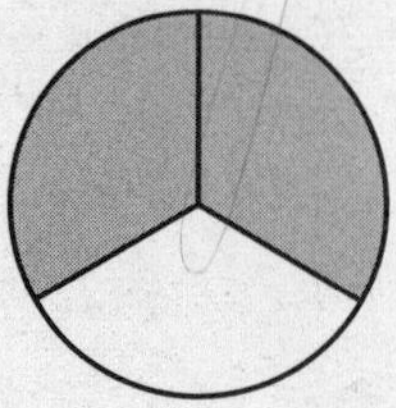

5. What fraction of the circle is shaded?

○ A. $\frac{7}{8}$

○ B. $\frac{5}{8}$

○ C. $\frac{3}{8}$

○ D. $\frac{1}{8}$

6. Where is point *Y* located on the number line?

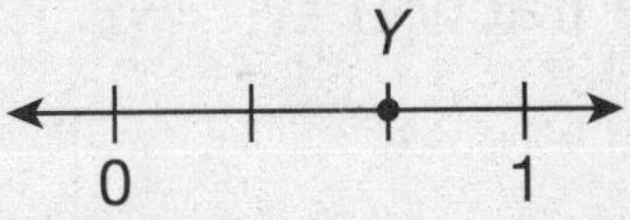

○ A. $\frac{1}{4}$

○ B. $\frac{1}{3}$

○ C. $\frac{1}{2}$

○ D. $\frac{2}{3}$

7. What fraction of the rectangle is shaded?

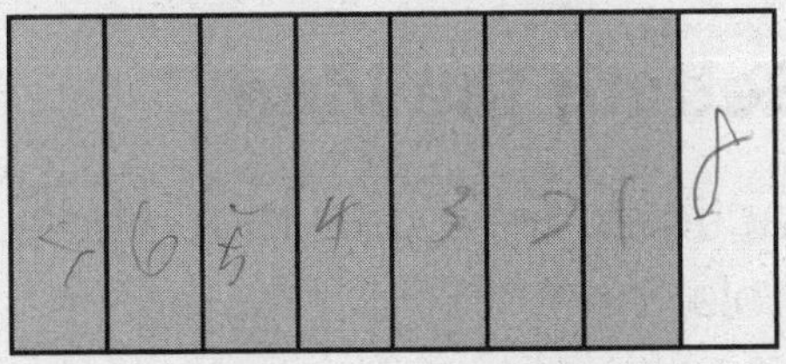

○ A. $\frac{7}{8}$

○ B. $\frac{6}{8}$

○ C. $\frac{5}{8}$

○ D. $\frac{1}{8}$

8. Where is point *H* located on the number line?

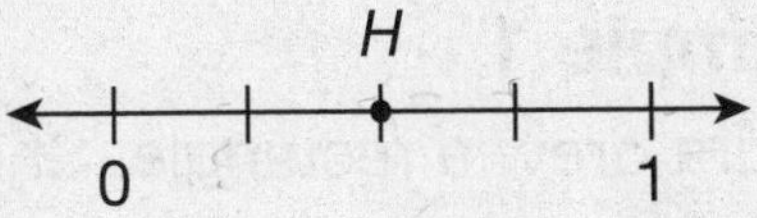

○ A. $\frac{1}{4}$

○ B. $\frac{2}{4}$

○ C. $\frac{2}{3}$

○ D. $\frac{3}{4}$

9. Lenny wants to show $\frac{3}{8}$ in two ways.

A. Shade the rectangle below to show $\frac{3}{8}$.

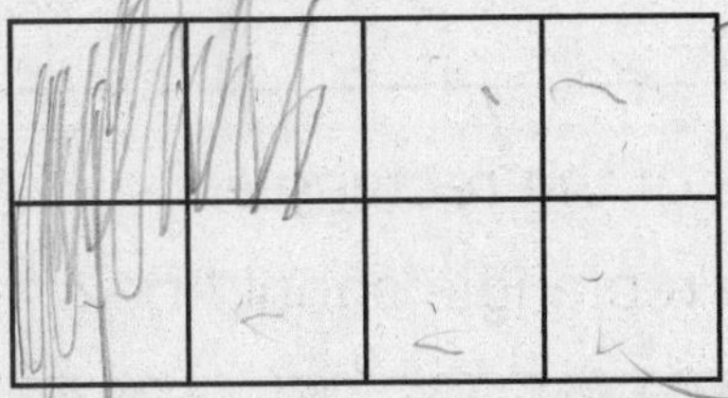

B. Draw point *R* at $\frac{3}{8}$ on the number line below.

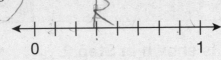

Common Core State Standard:
3.NF.3.c

Whole Numbers as Fractions

Getting the Idea

Fractions are a part of a whole. If you shade all parts, you can show a whole, or 1.

For example, the square below is divided into fourths.

Each part is $\frac{1}{4}$, and all four parts shaded show $\frac{4}{4}$.

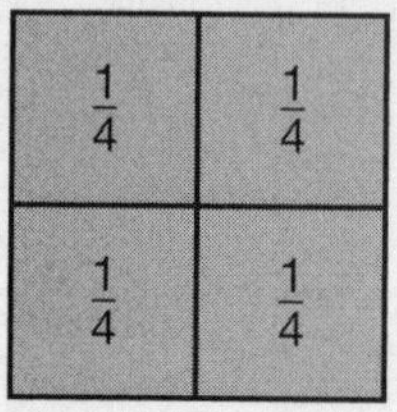

Example 1

Kendra drew a rectangle. She wants to show the fraction $\frac{6}{6}$.

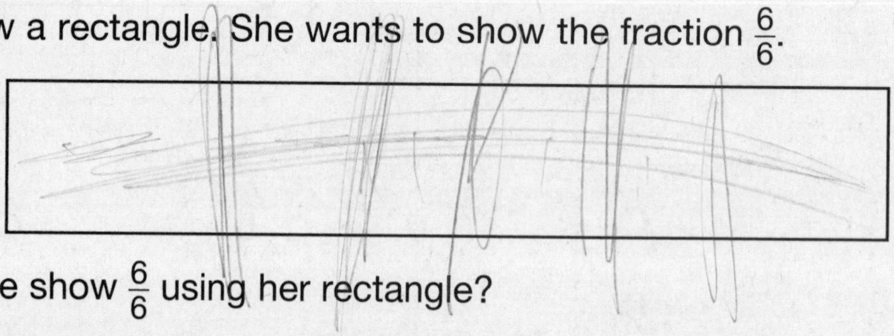

How can she show $\frac{6}{6}$ using her rectangle?

Strategy **Divide the rectangle into sixths. Shade 6 parts.**

Step 1 Divide Kendra's rectangle into 6 equal parts.

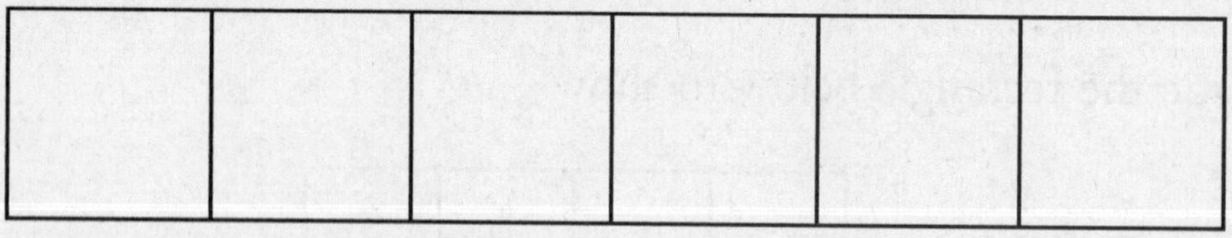

Step 2 Shade 6 parts of the rectangle.

The entire rectangle should be shaded since $\frac{6}{6} = 1$.

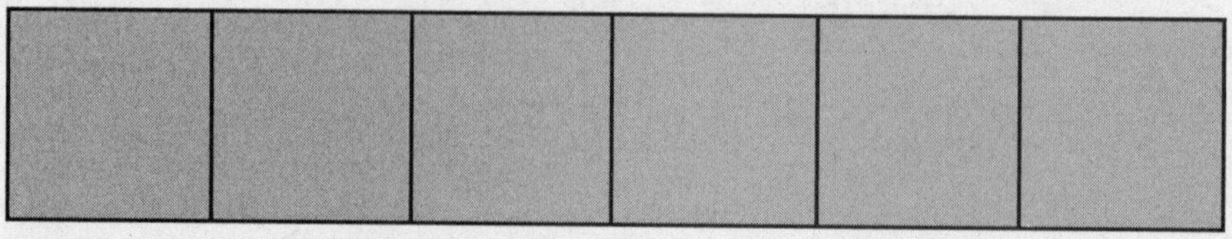

Solution **The answer is shown in Step 2.**

You can use a number line to show a whole. This number line is from 0 to 1.

There are four equal parts between 0 and 1. Each part represents $\frac{1}{4}$. So, $\frac{4}{4} = 1$.

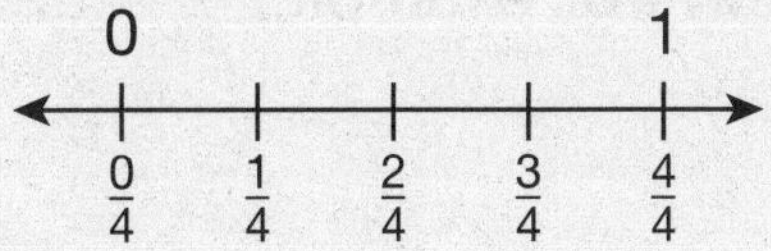

Example 2

How can you show $\frac{6}{6}$ as a whole on a number line?

Strategy **Draw a number line.**

Step 1 Draw a number line from 0 to 1.

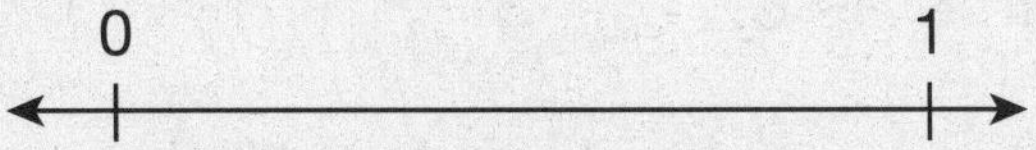

Step 2 Divide the number line into 6 equal parts.

Label each part.

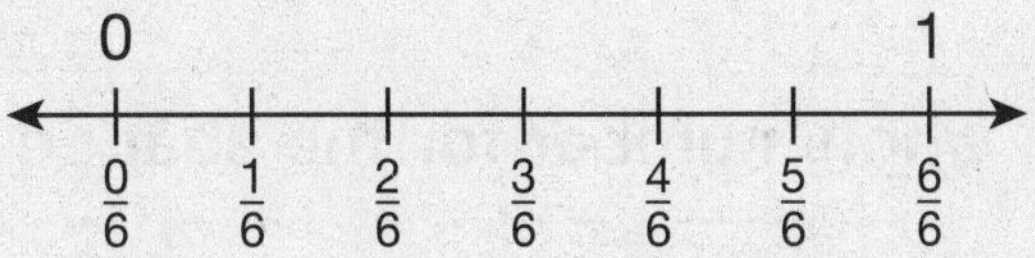

So, $\frac{6}{6} = 1$.

Solution **$\frac{6}{6}$ is shown as a whole on the number line in Step 2.**

You can show any whole number as a fraction.

When a whole number is the numerator and 1 is the denominator, the fraction is equal to the whole number. For example, $\frac{2}{1} = 2$.

Example 3

Mrs. Clark asked her class to write 8 as a fraction. What should her students write?

Strategy **Write the whole number over 1.**

Write 8 as the numerator.

Write 1 as the denominator.

$\frac{8}{1}$ is the same as 8.

Solution **Mrs. Clark's students should write $\frac{8}{1}$.**

Coached Example

Write a fraction and a whole number for the shaded part of the square below.

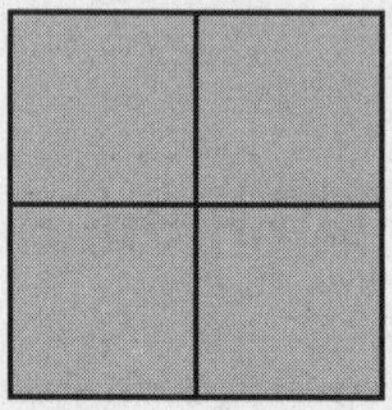

Write a fraction for each square.

How many equal parts are in the square? ________

How many equal parts are shaded? ________

What fraction does the rectangle show? ________

What whole number does the rectangle show? $\frac{4}{4}$ = ________

The fraction is ________ and the whole number is ________.

Lesson Practice

Choose the correct answer.

1. Which is the same as $\frac{3}{3}$?

○ **A.** 3

○ **B.** 2

○ **C.** 1

○ **D.** $\frac{1}{3}$

2. Which is another way to show the number 8?

○ **A.** $\frac{8}{1}$

○ **B.** $\frac{4}{4}$

○ **C.** $\frac{8}{8}$

○ **D.** $\frac{1}{8}$

3. Which fraction is equal to 1?

○ **A.** $\frac{2}{1}$

○ **B.** $\frac{2}{2}$

○ **C.** $\frac{2}{3}$

○ **D.** $\frac{2}{4}$

4. What fraction is shown by the picture below?

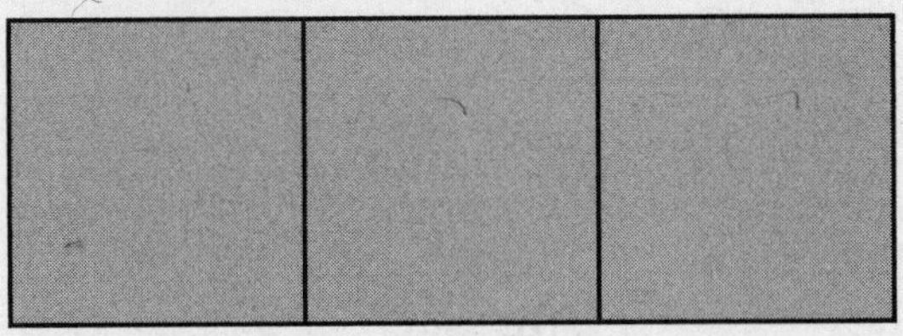

○ **A.** $\frac{1}{3}$

○ **B.** $\frac{2}{2}$

○ **C.** $\frac{3}{3}$

○ **D.** $\frac{3}{1}$

5. Which fraction is **not** equal to 1?

○ **A.** $\frac{3}{3}$

○ **B.** $\frac{5}{5}$

○ **C.** $\frac{7}{7}$

○ **D.** $\frac{9}{1}$

6. Which is another way to write the fraction $\frac{2}{2}$?

○ **A.** $\frac{1}{2}$

○ **B.** 1

○ **C.** 2

○ **D.** $\frac{2}{1}$

7. Which whole number is equal to $\frac{5}{1}$?

○ **A.** 1

○ **B.** 5

○ **C.** 6

○ **D.** 10

8. $\frac{10}{1} = \square$

○ **A.** 10

○ **B.** 9

○ **C.** 5

○ **D.** $\frac{1}{10}$

9. Mr. Torres asked his students to show $\frac{8}{8}$ with a rectangle and on a number line.

A. Shade the rectangle to show $\frac{8}{8}$.

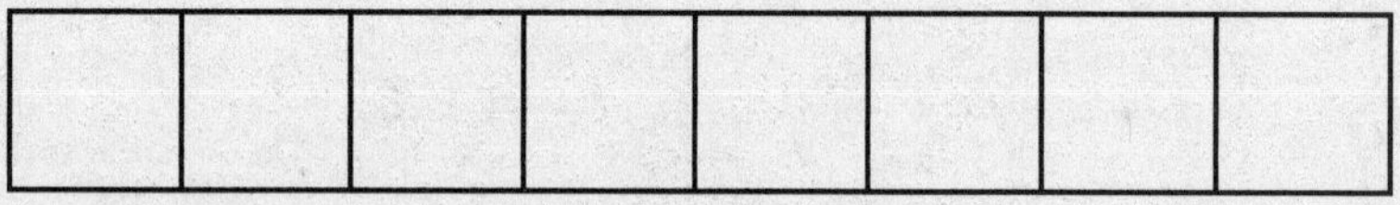

B. Label the number line. Draw point *A* at $\frac{8}{8}$.

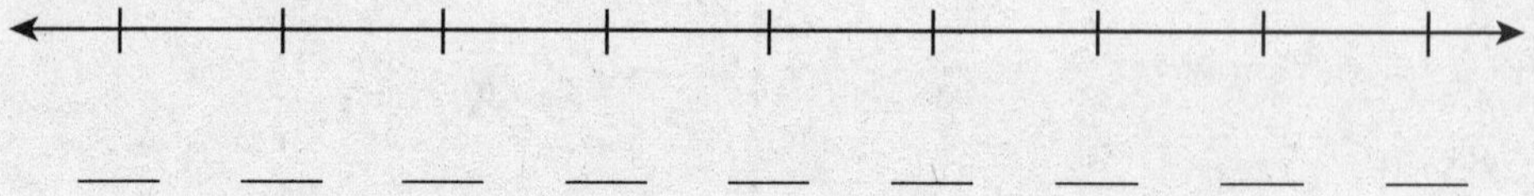

Common Core State Standards:
3.NF.3.a, 3.NF.3.b

Equivalent Fractions

Getting the Idea

Fractions can have different numerators and denominators and have the same value. These fractions name the same parts of a whole and are called **equivalent fractions**.

For example, the picture below shows the equivalent fractions $\frac{2}{3}$ and $\frac{4}{6}$.

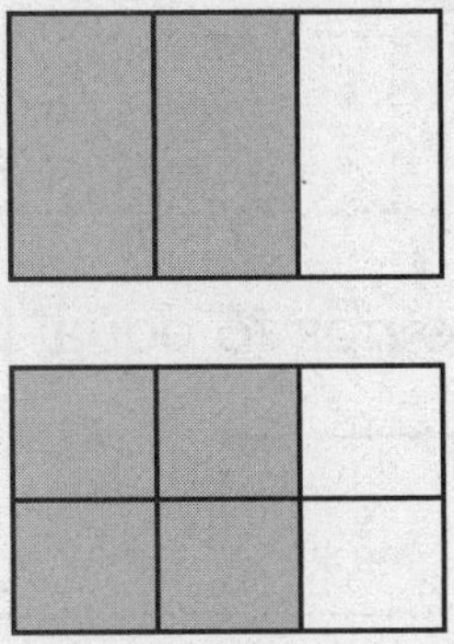

Example 1

Write two equivalent fractions that name the shaded parts of the circle.

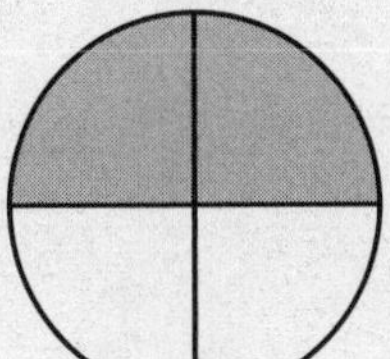

Strategy **Look at the shaded parts of the circle in two ways.**

Step 1 Count the number of equal parts and the number of shaded parts.

There are 4 equal parts. There are 2 shaded parts.

So, $\frac{2}{4}$ of the circle is shaded.

Step 2 Look at the shaded parts another way.

The two shaded parts are half of the circle.

One half of the circle is shaded.

So, $\frac{1}{2}$ of the circle is shaded.

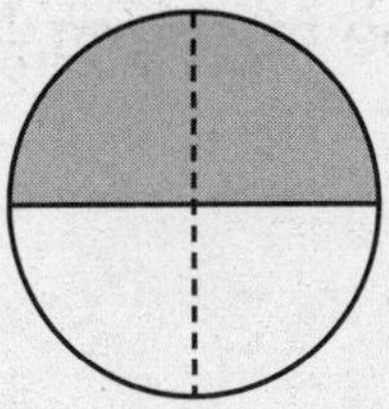

Solution **The fractions $\frac{1}{2}$ and $\frac{2}{4}$ are equivalent fractions that name the shaded parts of the circle.**

Example 2

Are $\frac{1}{2}$ and $\frac{4}{8}$ equivalent fractions?

Strategy **Use fraction strips.**

Step 1 Use the fraction strips for $\frac{1}{2}$ and $\frac{1}{8}$.

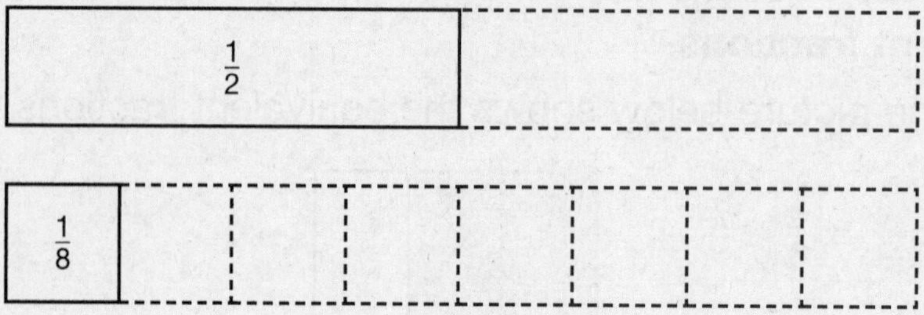

Step 2 Put together four $\frac{1}{8}$ strips to equal the length of the $\frac{1}{2}$ strip.

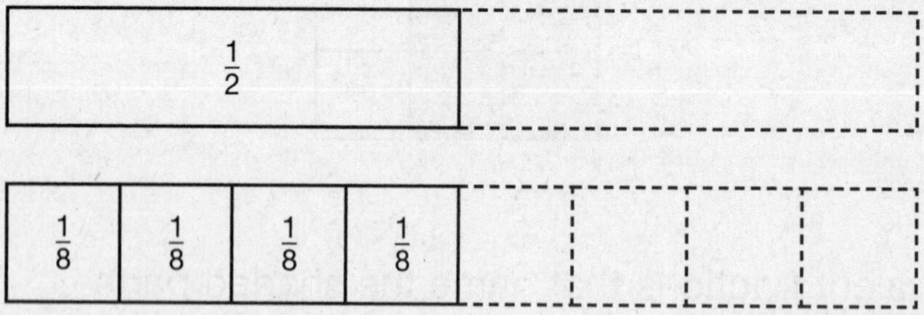

They are the same length, so the fractions are equivalent.

$\frac{1}{2} = \frac{4}{8}$

Solution **Yes, $\frac{1}{2}$ and $\frac{4}{8}$ are equivalent fractions.**

You can use a number line to find equivalent fractions.

The fractions $\frac{1}{2}$ and $\frac{3}{6}$ are at the same point on the number lines.

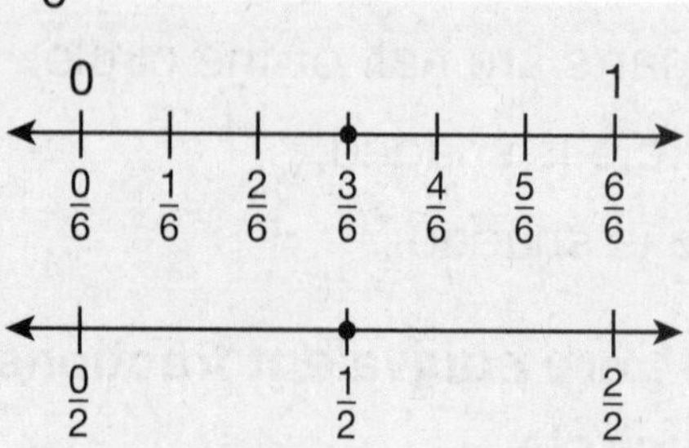

So, $\frac{1}{2} = \frac{3}{6}$.

Example 3

Rachael drew the number lines below.

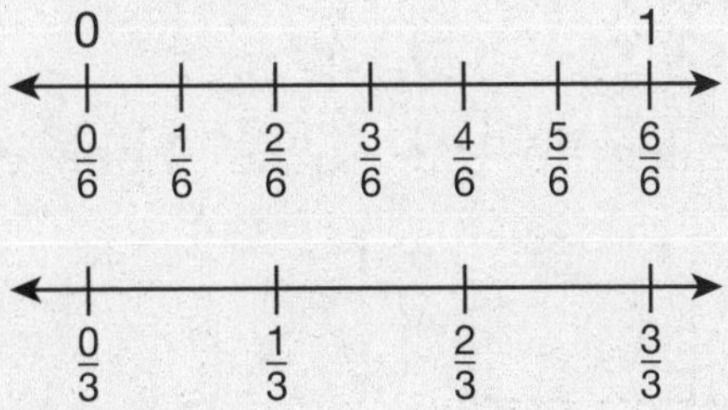

Which fraction is equivalent to $\frac{2}{6}$?

Strategy **Find the fraction at the same point on the number line as $\frac{2}{6}$.**

Find $\frac{2}{6}$ on the number line.

$\frac{1}{3}$ is at the same point as $\frac{2}{6}$.

So, $\frac{1}{3} = \frac{2}{6}$.

Solution **$\frac{1}{3}$ is equivalent to $\frac{2}{6}$.**

Coached Example

Evan thinks that $\frac{2}{4}$ and $\frac{4}{8}$ are equivalent fractions. Is Evan correct?

Draw number lines showing fourths and eighths.

Label the fourths. Then label the eighths.

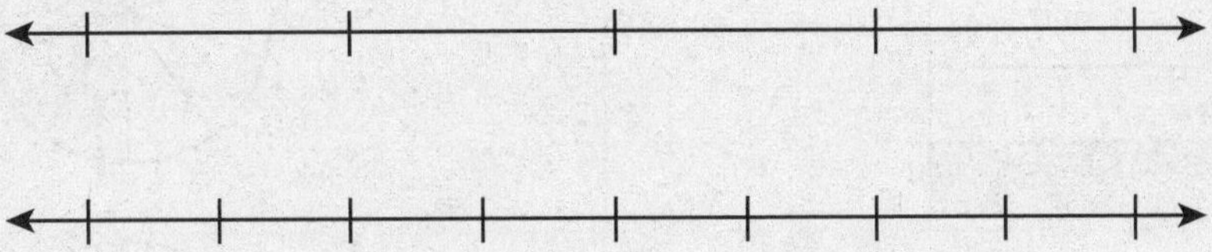

Are $\frac{2}{4}$ and $\frac{4}{8}$ at the same point on the number line? ____________________

Evan is ____________________________.

$\frac{2}{4}$ and $\frac{4}{8}$ ____________________________ equivalent fractions.

Lesson Practice

Choose the correct answer.

1. Which fraction is equivalent to $\frac{1}{4}$?

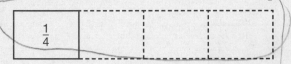

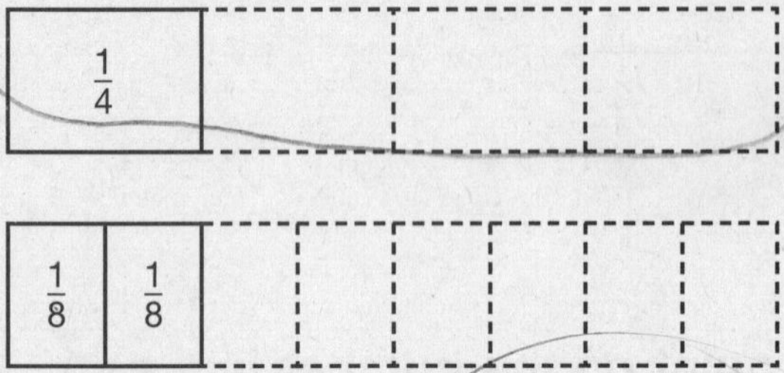

○ **A.** $\frac{1}{8}$

○ **B.** $\frac{2}{8}$

○ **C.** $\frac{4}{8}$

○ **D.** $\frac{6}{8}$

2. Which fraction is equivalent to $\frac{2}{6}$?

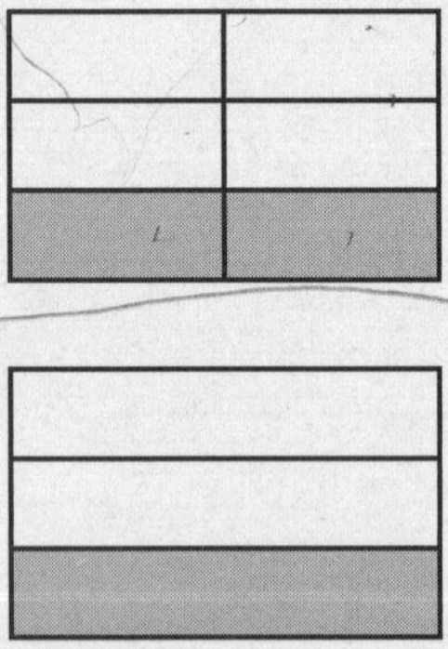

○ **A.** $\frac{1}{9}$

○ **B.** $\frac{1}{4}$

○ **C.** $\frac{1}{3}$

○ **D.** $\frac{1}{2}$

3. Look at the circle below.

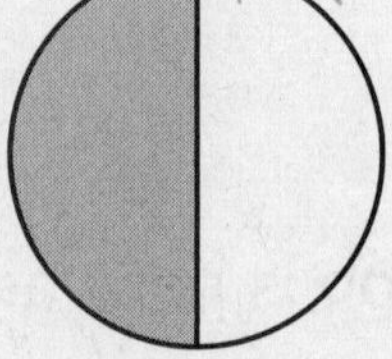

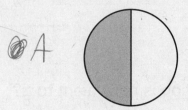

Which also shows $\frac{1}{2}$ of the circle shaded?

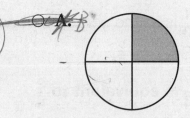

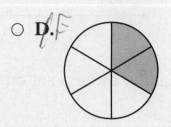

○ **A.**

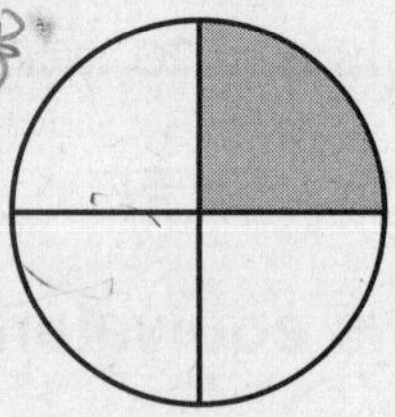

○ **B.**

○ **C.**

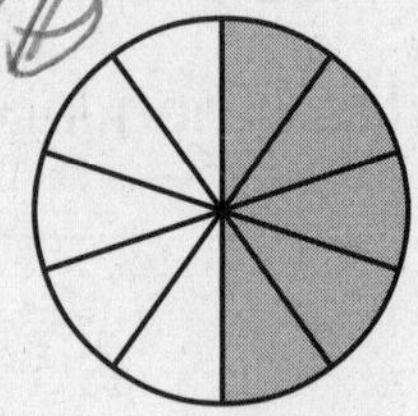

○ **D.**

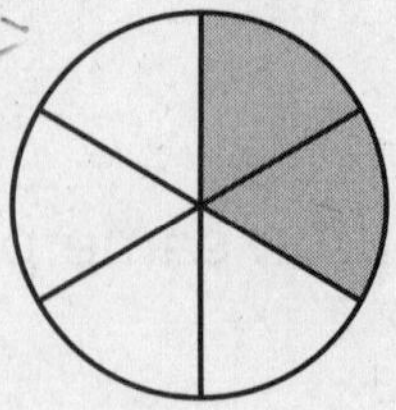

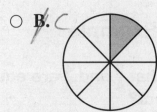

4. Which fraction is equivalent to $\frac{3}{4}$?

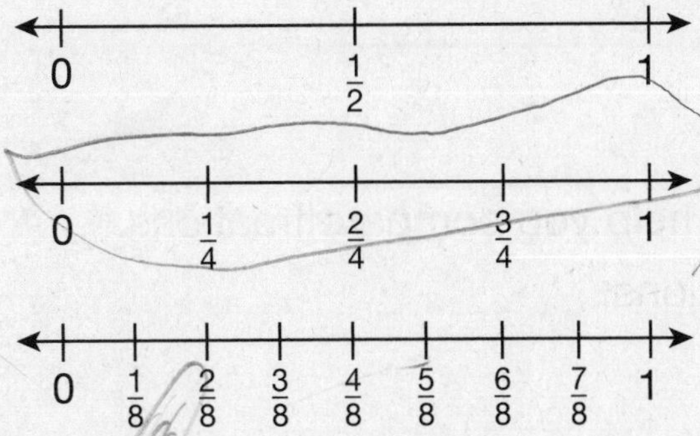

○ **A.** $\frac{2}{8}$

○ **B.** $\frac{3}{8}$

○ **C.** $\frac{1}{2}$

○ **D.** $\frac{6}{8}$

5. Which two fractions are equivalent?

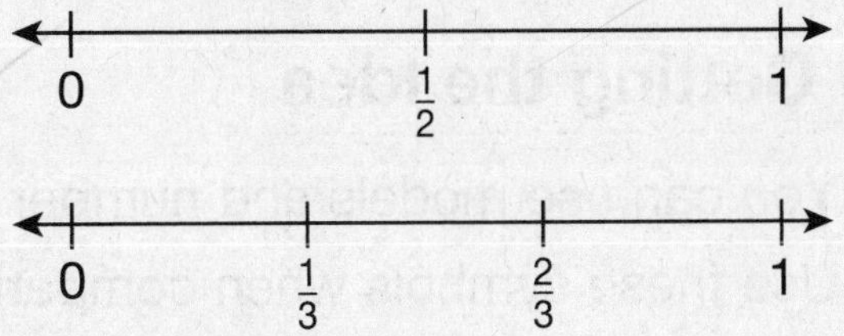

○ **A.** $\frac{2}{2}$ and $\frac{3}{3}$

○ **B.** $\frac{2}{2}$ and $\frac{2}{3}$

○ **C.** $\frac{1}{2}$ and $\frac{2}{3}$

○ **D.** $\frac{1}{2}$ and $\frac{1}{3}$

6. Look at the models below.

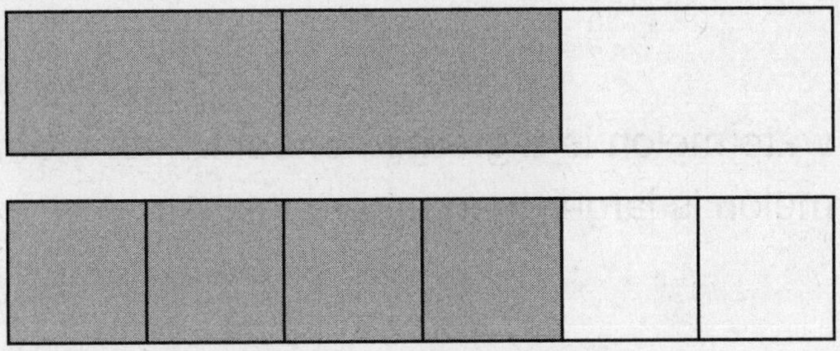

A. Write two equivalent fractions for the models.

________________ and ________________

B. Show the two equivalent fractions on the number lines below.

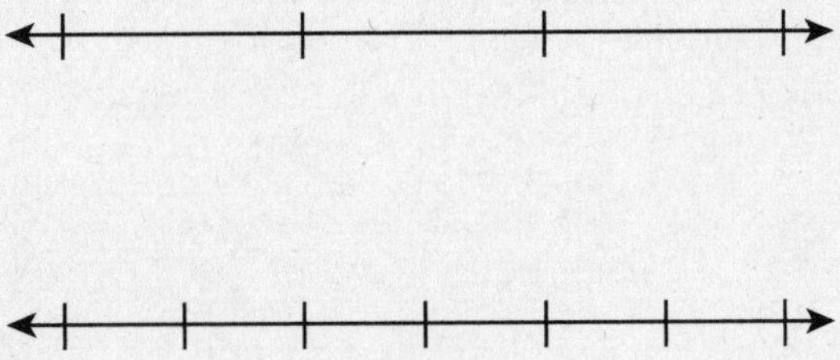

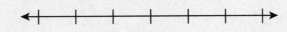

Compare Fractions

Common Core State Standard:
3.NF.3.d

Getting the Idea

You can use models and number lines to help you compare fractions.

Use these symbols when comparing fractions.

> means **is greater than.**

< means **is less than.**

= means **is equal to.**

You can compare fractions that have the same numerator or the same denominator.

When you compare, it is important that the wholes are the same size. For example, $\frac{1}{3}$ is less than $\frac{1}{2}$ as is shown on the number lines below.

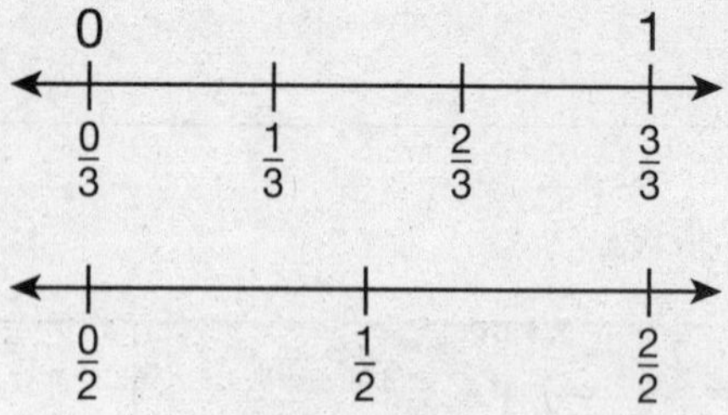

However, $\frac{1}{3}$ of a watermelon is a greater amount than $\frac{1}{2}$ of an orange, because a watermelon is larger than an orange.

Example 1

Which symbol makes this sentence true? Write >, <, or =.

$\frac{5}{8} \bigcirc \frac{3}{8}$

Strategy **Use fraction strips to compare $\frac{5}{8}$ and $\frac{3}{8}$.**

Step 1 Show $\frac{5}{8}$ and $\frac{3}{8}$ with fraction strips.

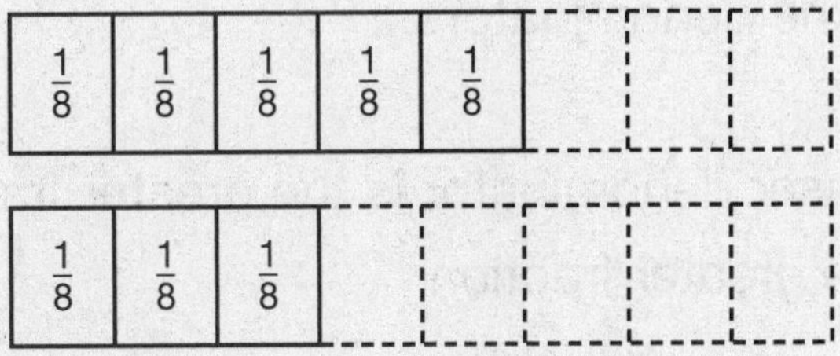

Step 2 Compare the fractions.

$\frac{5}{8}$ has 5 parts shaded. $\frac{3}{8}$ has 3 parts shaded.

5 parts is more than 3 parts.

So, $\frac{5}{8}$ is greater than $\frac{3}{8}$.

Solution $\frac{5}{8}$ (>) $\frac{3}{8}$

When you compare fractions with the same denominators, the fraction with the greater numerator is the greater fraction.
In Example 1, 5 is greater than 3, so $\frac{5}{8}$ is greater than $\frac{3}{8}$.

When you compare fractions with the same numerators, the fraction with the lesser denominator is the greater fraction.
For example, all the fractions shown below have 1 as their numerator.
As the number of equal parts (the denominator) increases, the size of each part decreases.

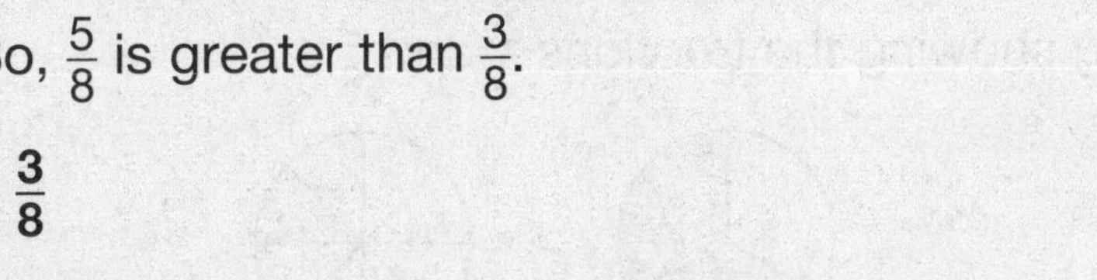

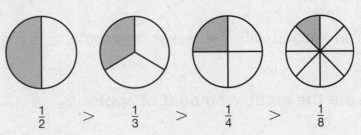

$\frac{1}{2} > \frac{1}{3} > \frac{1}{4} > \frac{1}{8}$

Example 2

Which symbol makes this sentence true? Write >, <, or =.

$\frac{2}{3} \bigcirc \frac{2}{8}$

Strategy **Compare the denominators.**

Step 1 The numerators are the same.

Step 2 Compare the denominators.

$3 < 8$

The lesser denominator is the greater fraction.

$\frac{2}{3}$ is the greater fraction.

Step 3 Choose the correct symbol.

> means is greater than.

Solution $\frac{2}{3} \circled{>} \frac{2}{8}$

Here is a drawing showing the fractions $\frac{2}{3}$ and $\frac{2}{8}$.

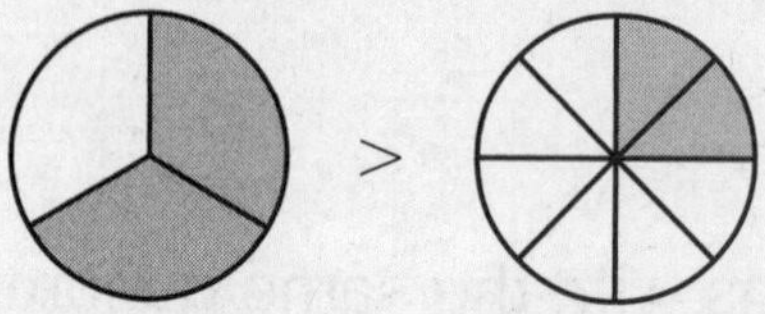

Example 3

Ted ate $\frac{2}{4}$ of his apple. Lisa ate $\frac{2}{3}$ of her apple. Who ate the greater amount?

Strategy **Compare the denominators.**

Step 1 Both of the numerators are 2.

Step 2 Compare the denominators.

$4 > 3$

The fraction with the lesser denominator is the greater fraction.

$\frac{2}{3} > \frac{2}{4}$

Solution **Lisa ate the greater amount of apple.**

Here are the fractions from Example 3 on number lines.

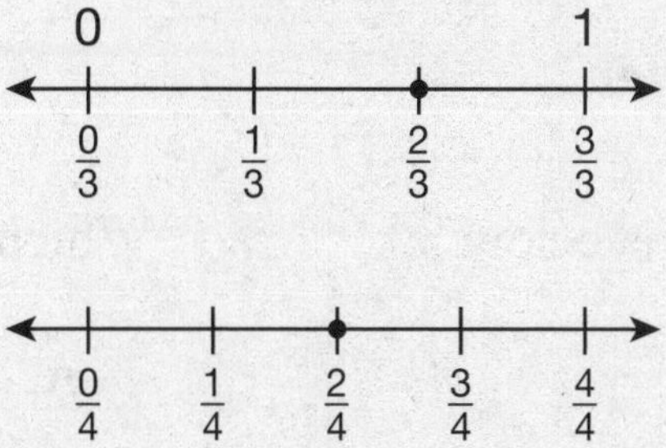

The fraction farther to the right is the greater fraction.

The fraction farther to the left is the lesser fraction.

Coached Example

Callie drew a circle. Callie shaded $\frac{1}{2}$ of the circle. Will shaded $\frac{1}{4}$ of the circle. Who shaded more of the circle?

Draw number lines divided into halves and fourths.
Draw points at $\frac{1}{2}$ and $\frac{1}{4}$ on the number lines.

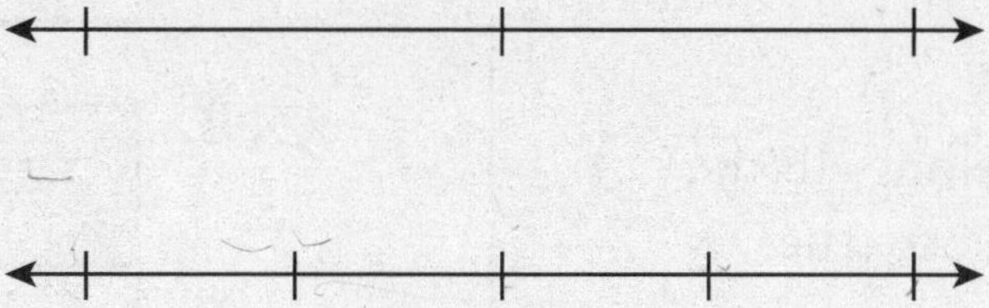

Compare the fractions.

The fraction farther to the right is the ______________ fraction.

So, ____________ > ____________.

______________ **shaded more of the circle than** ______________.

Lesson Practice

Choose the correct answer.

1. Look at the two fractions below.

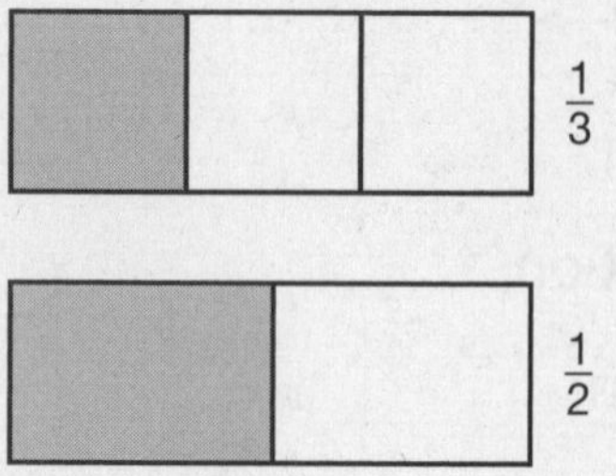

Which sentence is true?

○ **A.** $\frac{1}{3} = \frac{1}{2}$

○ **B.** $\frac{1}{3} > \frac{1}{2}$

○ **C.** $\frac{1}{2} < \frac{1}{3}$

○ **D.** $\frac{1}{2} > \frac{1}{3}$

2. Which symbol belongs in the ○ to make the sentence true?

$\frac{4}{6}$ ○ $\frac{2}{6}$

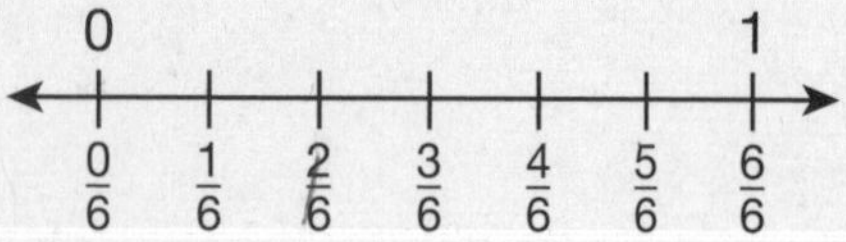

○ **A.** >

○ **B.** <

○ **C.** =

○ **D.** +

3. Which is the least fraction?

○ **A.** $\frac{1}{8}$

○ **B.** $\frac{2}{8}$

○ **C.** $\frac{3}{8}$

○ **D.** $\frac{4}{8}$

4. Which is the greatest fraction?

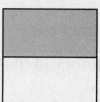

○ **A.**

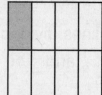

○ **B.**

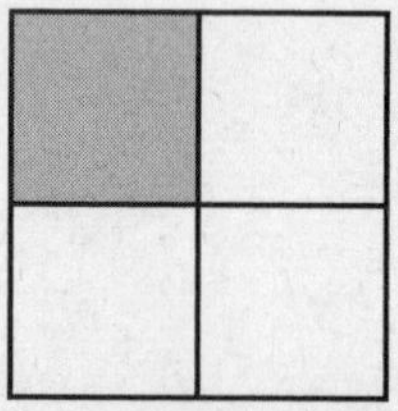

○ **C.**

○ **D.**

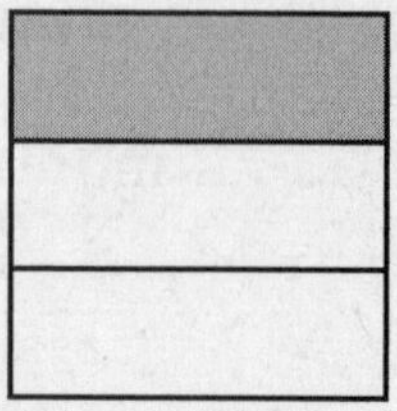

5. The circle below is $\frac{1}{4}$ shaded.

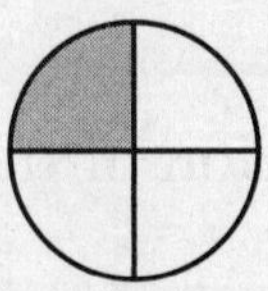

Which circle has less than $\frac{1}{4}$ shaded?

○ **A.**

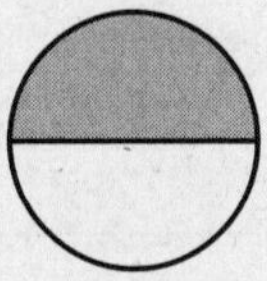

○ **B.**

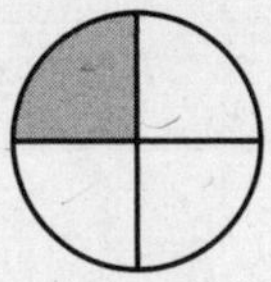

○ **C.**

○ **D.**

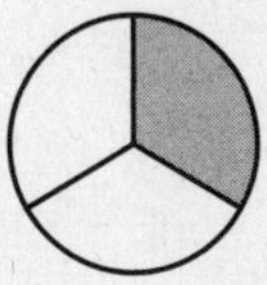

6. Which symbol belongs in the ○ to make the sentence true?

$\frac{3}{4}$ ○ $\frac{3}{8}$

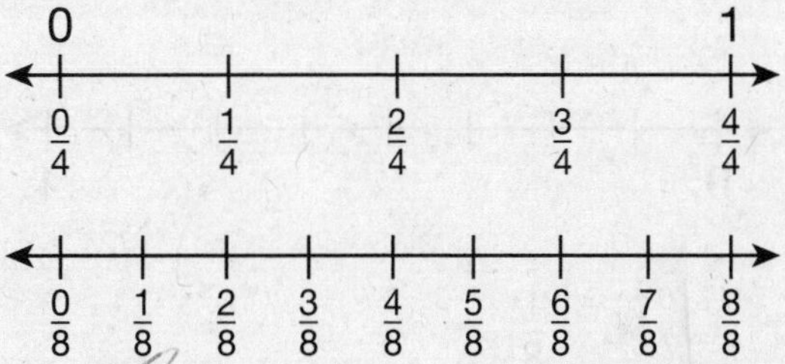

○ **A.** >

○ **B.** <

○ **C.** =

○ **D.** +

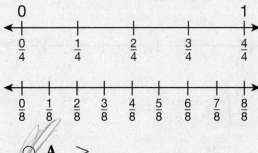

7. Brenda has read $\frac{5}{6}$ of a book. Sylvia has read $\frac{5}{8}$ of the same book.

A. Circle $\frac{5}{6}$ and $\frac{5}{8}$ on the number lines below.

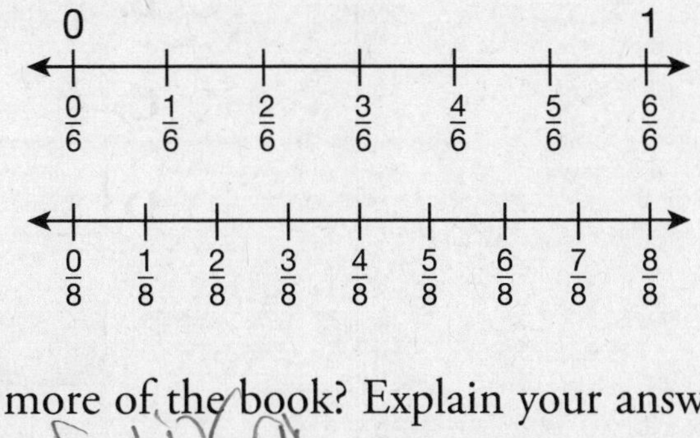

B. Who has read more of the book? Explain your answer.

Domain 3: Cumulative Assessment for Lessons 19–22

1. Where is point *A* located on the number line?

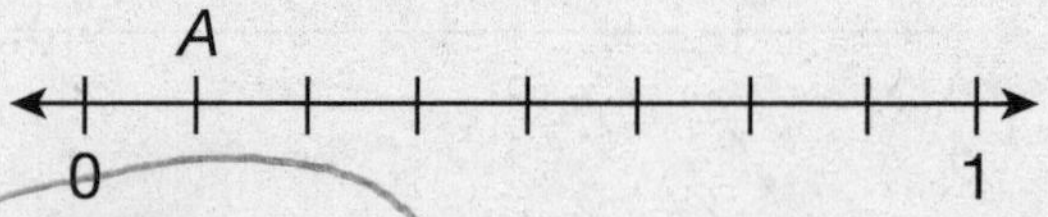

○ **A.** $\frac{1}{8}$

○ **B.** $\frac{3}{8}$

○ **C.** $\frac{5}{8}$

○ **D.** $\frac{7}{8}$

2. Which two fractions are equivalent?

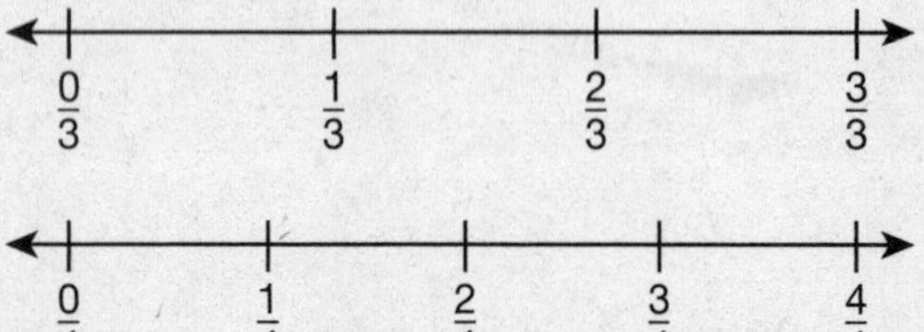

○ **A.** $\frac{1}{3}$ and $\frac{1}{4}$

○ **B.** $\frac{1}{3}$ and $\frac{2}{4}$

○ **C.** $\frac{2}{3}$ and $\frac{3}{4}$

○ **D.** $\frac{3}{3}$ and $\frac{4}{4}$

3. What fraction of the rectangle is shaded?

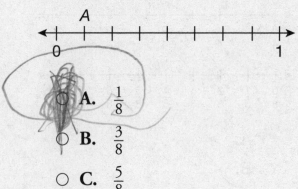

○ **A.** $\frac{3}{6}$

○ **B.** $\frac{4}{6}$

○ **C.** $\frac{5}{6}$

○ **D.** $\frac{6}{6}$

4. Which is the same as $\frac{8}{8}$?

○ **A.** 1

○ **B.** $\frac{1}{8}$

○ **C.** $\frac{8}{1}$

○ **D.** 8

5. Which symbol belongs in the ◯ to make the sentence true?

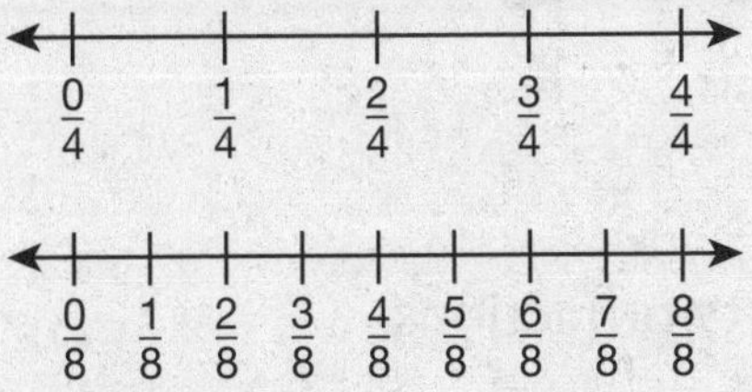

$\frac{1}{8}$ ◯ $\frac{1}{4}$

- ○ **A.** >
- ○ **B.** <
- ○ **C.** =
- ○ **D.** +

6. Which fraction is equivalent to $\frac{4}{6}$?

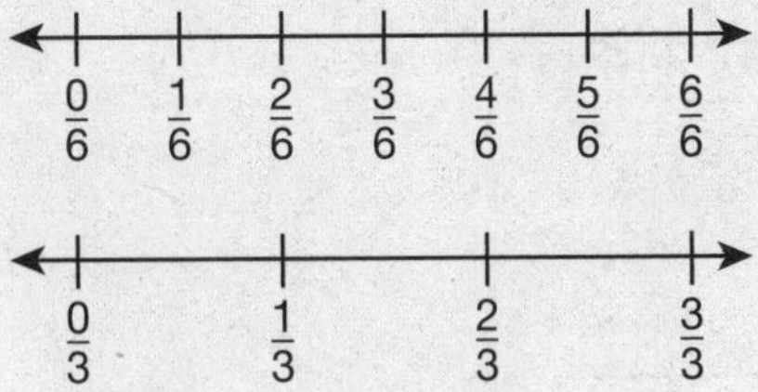

- ○ **A.** $\frac{2}{3}$
- ○ **B.** $\frac{1}{3}$
- ○ **C.** $\frac{1}{4}$
- ○ **D.** $\frac{1}{6}$

7. Which symbol belongs in the ◯ to make the sentence true?

$\frac{3}{4}$ ◯ $\frac{1}{4}$

- ○ **A.** >
- ○ **B.** <
- ○ **C.** =
- ○ **D.** +

8. Which is another way to represent the number 6?

- ○ **A.** $\frac{1}{6}$
- ○ **B.** $\frac{6}{6}$
- ○ **C.** $\frac{6}{1}$
- ○ **D.** $\frac{9}{1}$

9. Draw point C at $\frac{5}{8}$ on the number line below.

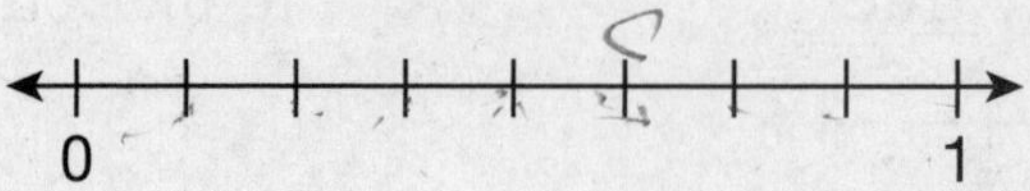

10. Sergio made a drawing of two circles to show two fractions.

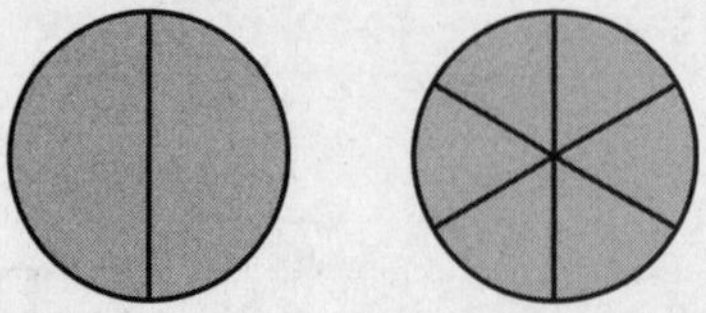

A. Write two equivalent fractions for Sergio's models.

____________________ and ____________________

B. Show the two equivalent fractions on the number lines below.

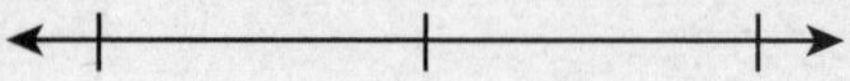

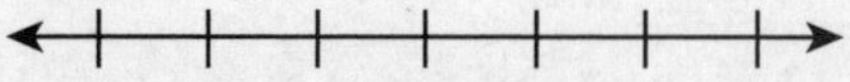

Domain 4

Measurement and Data

Domain 4: Diagnostic Assessment for Lessons 23–33

1. The clock shows the time the game started.

What time did the game start?

○ **A.** 7:07 ○ **C.** 8:07

○ **B.** 7:10 ○ **D.** 8:10

2. Which most likely has a mass of about 2 grams?

○ **A.**

○ **B.**

○ **C.**

○ **D.**

3. What is the perimeter of this shape?

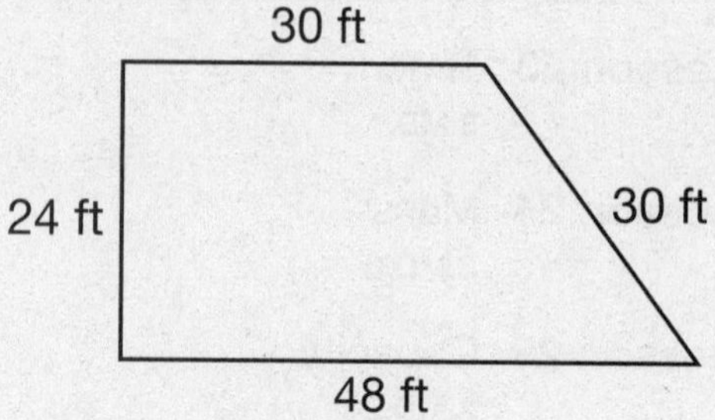

○ **A.** 78 feet

○ **B.** 98 feet

○ **C.** 112 feet

○ **D.** 132 feet

4. What is the area of the shaded rectangle?

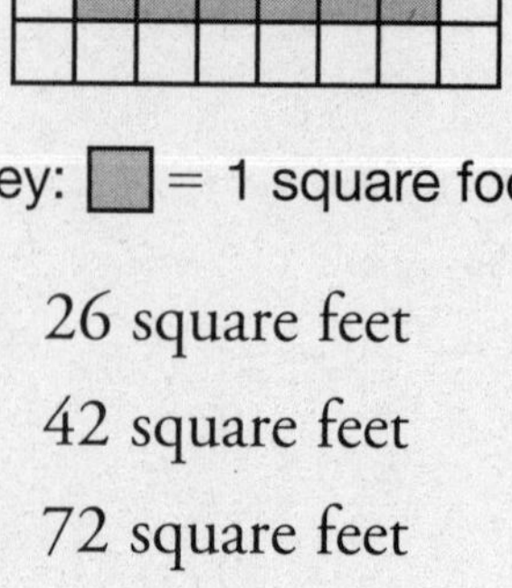

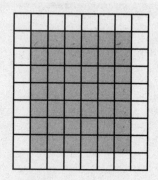

Key: ■ = 1 square foot

○ **A.** 26 square feet

○ **B.** 42 square feet

○ **C.** 72 square feet

○ **D.** 76 square feet

5. What is the area of the rectangle?

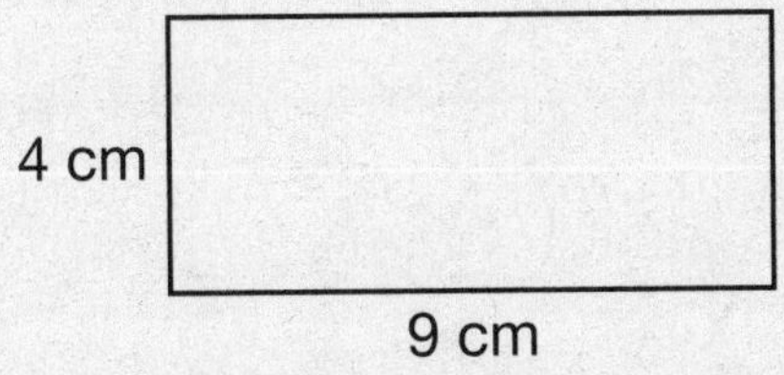

4 cm

9 cm

○ **A.** 26 square centimeters

○ **B.** 27 square centimeters

⊘ **C.** 36 square centimeters

○ **D.** 40 square centimeters

6. The picture graph shows the number of CDs three students own.

CDs Owned

Jay	◉ ◉ ◉ ◉ ◉ ◉ ◉
Pete	◉ ◉ ◉ ◉ ◉ ◉
Ben	◉ ◉ ◉ ◉ ◉

Key: Each ◉ = 2 CDs

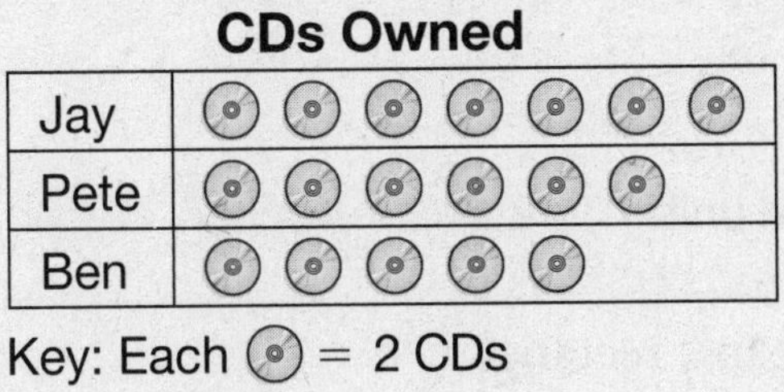

How many more CDs does Pete own than Ben?

○ **A.** 1

⊘ **B.** 2

○ **C.** 5

○ **D.** 6

7. Dwayne made 800 milliliters of tea in a pot. He poured 145 milliliters into a cup. How much tea is left in the pot?

○ **A.** 655 milliliters

○ **B.** 665 milliliters

○ **C.** 700 milliliters

⊘ **D.** 945 milliliters

8. What is the area of the figure below?

Key: □ = 1 square foot

○ **A.** 12 square feet

○ **B.** 24 square feet

○ **C.** 56 square feet

⊘ **D.** 60 square feet

9. The line plot shows the lengths of some ants.

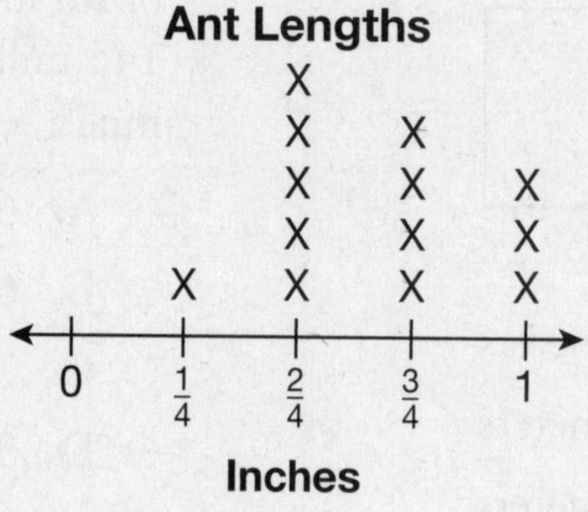

How many ants are $\frac{3}{4}$ inch or longer?

---

10. Megan made this rectangle.

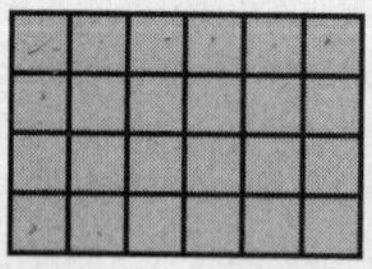

Key: ▢ = 1 square unit

A. What are the perimeter and area of Megan's rectangle?

B. On the grid below, draw a rectangle with the same area but a different perimeter than Megan's rectangle.

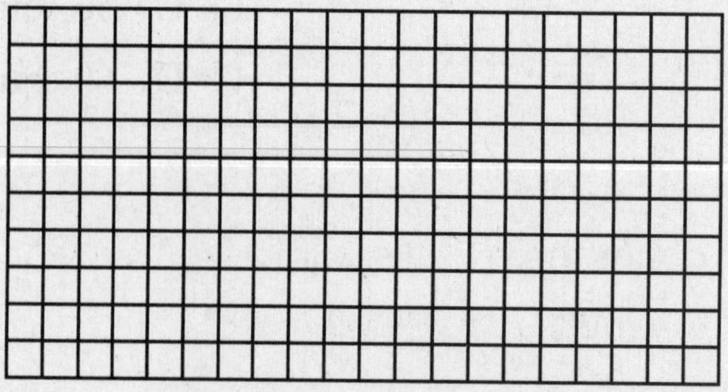

Key: ▢ = 1 square unit

Time

Common Core State Standard:
3.MD.1

Getting the Idea

A clock is used to tell and measure time.

On an analog clock, the short hand points to the **hour**.
The long hand points to the **minute**.

The minute hand moves to the next number every 5 minutes.

The hour hand moves to the next number every 60 minutes or 1 hour.

Example 1

The clock shows the time that Aiden's bus picks him up for school.

At what time does Aiden's bus pick him up?

Strategy **Look at the hands of the clock.**

Step 1 Look at the shorter hand to tell the hour.

The hand is between 8 and 9, so it is after 8 o'clock.

Step 2 Look at the longer hand to tell the minutes.

The minute hand is between 2 and 3.

Skip count the minutes by 5s from 8:00 to 8:10.

Then count the minutes by 1s from 8:10 to the minute hand.

It is 12 minutes past the hour.

Solution **Aiden's bus picks him up at 8:12.**
Read: eight twelve or twelve minutes past eight o'clock

Example 2

The clock shows the time that Margie got home from school today.

At what time did Margie get home from school?

Strategy **Look at the hands of the clock.**

Step 1 Look at the hour hand.

It is between 3 and 4, so it is after 3 o'clock.

Step 2 Look at the minute hand.

When the minute hand points to the 7, it is 35 minutes past the hour.

The minute hand points to the second mark after 7.

Count forward 2 minutes from 35.

It is 37 minutes past the hour.

Solution **Margie got home at 3:37 today.**
Read: three thirty-seven or thirty-seven minutes past three o'clock

Midnight is 12:00 A.M. Noon is 12:00 P.M.
The hours between midnight and noon are called **A.M.**
The hours between noon and midnight are called **P.M.**

Elapsed time is the amount of time from the start to the finish of an event.
For example, you started a quiz at 10:04 A.M. and finished the quiz at 10:15 A.M.
The elapsed time is 11 minutes.

Example 3

Regina started writing in her journal at 4:15 P.M. She finished writing at 4:45 P.M. How much time was Regina writing in her journal?

Strategy **Use an analog clock. Skip count.**

Step 1 Show 4:15 on the clock. Then move the minute hand to 4:45.

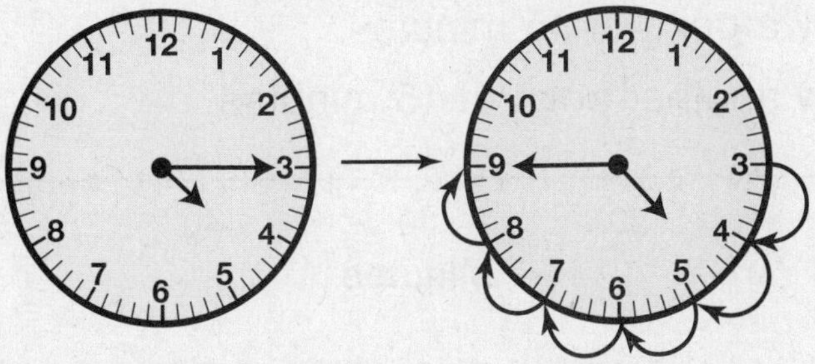

Step 2 Skip count by 5s.

5 → 10 → 15 → 20 → 25 → 30

So, 30 minutes, or $\frac{1}{2}$ hour, has passed.

Solution **Regina was writing in her journal for 30 minutes, or $\frac{1}{2}$ hour.**

You can use a number line to help you add or subtract to solve word problems involving time.

Last night, Rose weeded her garden for 15 minutes. Then she watered for 18 minutes. How much time did she tend to her garden? Show the problem on a number line.

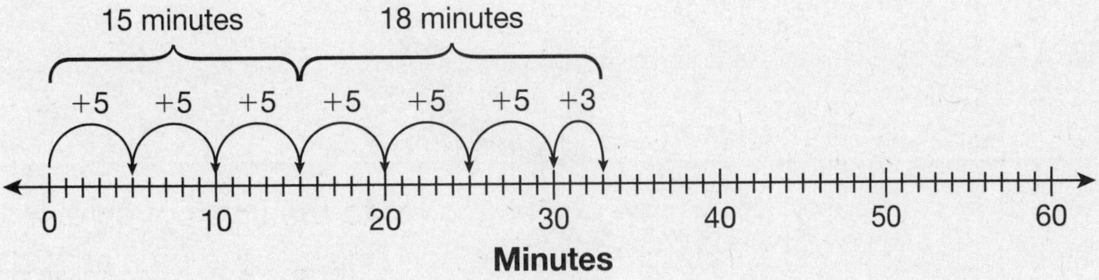

15 minutes + 18 minutes = 33 minutes

Rose tended to her garden for 33 minutes last night.

Example 4

Nia studied 45 minutes for her spelling test. Mike studied 15 minutes for the same test. How much longer did Nia study than Mike?

Strategy **Use a number line.**

Step 1 Make a number line from 0 to 60.

Draw a point at 45 minutes.

Draw another point at 15 minutes.

Step 2 Count back or subtract to find the difference in times.

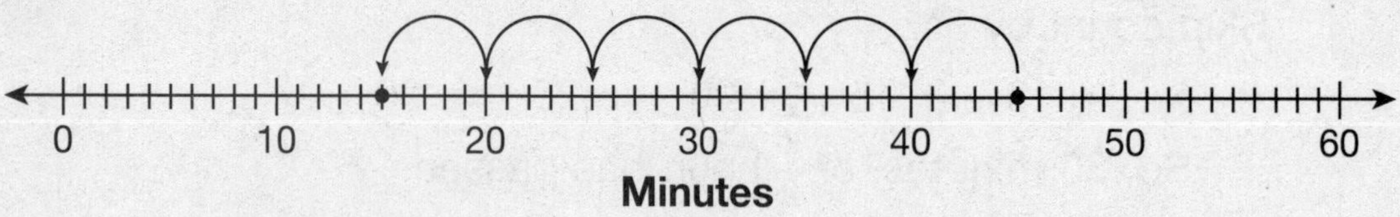

45 minutes − 15 minutes = 30 minutes

Solution **Nia studied 30 minutes longer for the math test than Mike.**

Coached Example

Alyssa got up at 7:10 A.M. She has to be ready by 7:55 A.M. for the bus. How much time does Alyssa have to get ready before the bus arrives?

Use a number line to find the time difference.

7:10 A.M. is ________ minutes after 7, so make a point at ________ minutes.

7:55 A.M. is ________ minutes after 7, so make another point at ________ minutes.

Find the time difference.

Count back on the number line or subtract.

55 minutes − 10 minutes = __________ minutes

Alyssa has ________ minutes to get ready before the bus arrives.

Lesson Practice

Choose the correct answer.

1. The clock shows the time that Quincy finished football practice.

What time did Quincy finish football practice?

○ **A.** 5:28
○ **B.** 5:43
○ **C.** 6:43
○ **D.** 8:28

2. Willa's ballet class starts at 9:15. Which is **not** a way to read the time when ballet class starts?

○ **A.** nine fifteen
○ **B.** half past nine
○ **C.** a quarter past nine
○ **D.** fifteen minutes after nine

3. Glenn started studying at 12:23 P.M. He finished at 12:58 P.M. How long did Glenn study?

○ **A.** 5 minutes
○ **B.** 15 minutes
○ **C.** 25 minutes
○ **D.** 35 minutes

Use the clock below for questions 4 and 5.

The clock shows the time that Ed got on the subway for his trip to the dentist's office.

4. What time did Ed get on the subway?

○ **A.** 8:10
○ **B.** 8:01
○ **C.** 9:10
○ **D.** 9:30

5. Ed rode the subway for 25 minutes. Then he walked for 13 minutes to get to the office. How long was Ed's trip to the dentist?

○ **A.** 38 minutes
○ **B.** 28 minutes
○ **C.** 25 minutes
○ **D.** 12 minutes

Use the number line to help you solve questions 6 and 7.

6. Before dinner, Alicia watched a show for 30 minutes. Then she played with her dog for 17 minutes. How much longer did Alicia watch the show than play with her dog?

○ **A.** 13 minutes ○ **C.** 23 minutes

○ **B.** 18 minutes ○ **D.** 47 minutes

7. For his daily exercise, Howie ran for 22 minutes. Then he walked for 5 minutes. How many minutes did he exercise in all?

○ **A.** 11 minutes ○ **C.** 17 minutes

○ **B.** 15 minutes ○ **D.** 27 minutes

8. The clock shows the time some students have gym class.

A. What time do the students have gym class? Include A.M. or P.M.

1:00

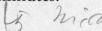

B. During gym class the students did sit-ups for 3 minutes. Then they danced for 42 minutes.

How long is gym class, in minutes?

45 min

Common Core State Standard:
3.MD.2

Mass

Getting the Idea

Mass is the measure of the amount of matter in an object.
The more mass an object has, the heavier it is.
The table shows two units of mass in the metric system.

Metric Units of Mass
1 **kilogram (kg)** = 1,000 **grams (g)**

Use these benchmarks to estimate mass.

A paper clip has a mass of about 1 gram.

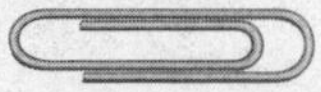

A pair of sneakers has a mass of about 1 kilogram.

To measure the mass of an object, you can use a **scale.**

Below are some different scales.

You can measure mass using grams or kilograms.

Example 1

Each mass on the right side of the scale is 1 gram.

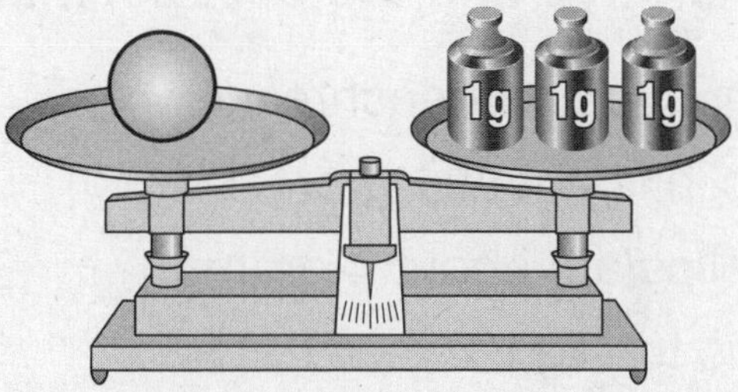

What is the mass of the ping-pong ball?

Strategy **Use a balance scale. Experiment with gram masses until the trays are even.**

Step 1 Make sure the scale is balanced.

When the scale is balanced, the mass is equal on both sides.

Step 2 Find the total amount of mass on the right side of the scale.

There are 3 masses.

Each is 1 gram.

So, the total mass is 3 grams.

Solution **The ping-pong ball has a mass of about 3 grams.**

Example 2

Which is the better estimate for the mass of a laptop computer?

4 grams   4 kilograms

Strategy **Use benchmarks to find the best estimate.**

Step 1 Compare 4 grams to a benchmark.

1 paper clip is about 1 gram.

4 paper clips are about 4 grams.

4 grams is too light.

Step 2 Compare 4 kilograms to a benchmark.

A pair of sneakers is about 1 kilogram.

4 pairs of sneakers are about 4 kilograms.

4 kilograms is about right for the mass of a laptop.

Solution **A laptop computer has a mass of about 4 kilograms.**

Example 3

Tyson, a Rottweiler, has a mass of 43 kilograms. Louie, a bulldog, is 18 kilograms lighter than Tyson. What is Louie's mass, in kilograms?

Strategy **Look for key words to decide which operation to use.**

Step 1 Decide which operation to use.

"18 kilograms lighter than Tyson" means to subtract.

Find $43 - 18 = \square$.

Step 2 Subtract.

$$\begin{array}{r} 43 \\ -\ 18 \\ \hline 25 \end{array}$$

Step 3 Use addition to check your answer.

$25 + 18 = 43$

The answer is correct.

Solution **Louie has a mass of 25 kilograms.**

Example 4

A cherry has a mass of 7 grams. Pat ate 9 cherries.
How many grams of cherries did Pat eat in all?

Strategy **Look for key words to decide which operation to use.**

Step 1 Decide which operation to use.

"How many grams in all" means to add or multiply.

Multiplication will be faster.

Find $9 \times 7 = \square$.

Step 2 Multiply.

$9 \times 7 = 63$

Solution **Pat ate 63 grams of cherries in all.**

Coached Example

This watermelon has a mass of 20 kilograms.

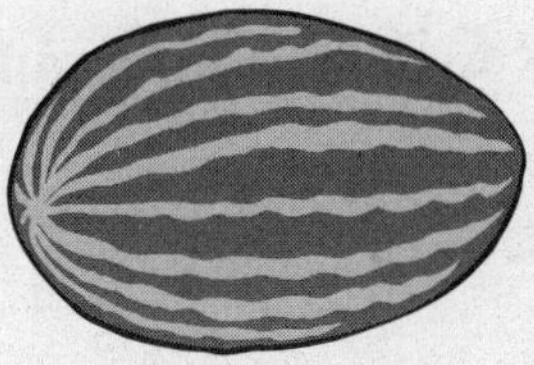

**Mr. Lopez cut the watermelon into 5 equal pieces.
What is the mass, in kilograms, of each piece?**

Decide which operation to use.

The watermelon was cut into ________ equal pieces.

To find the mass of each piece, use ___________________.

Find 20 ÷ ________ = □.

Divide.

________ ÷ ________ = ________

The mass of each piece is __________ kilograms.

Lesson Practice

Choose the correct answer.

1. Each mass on the right side of the scale is 1 kilogram.

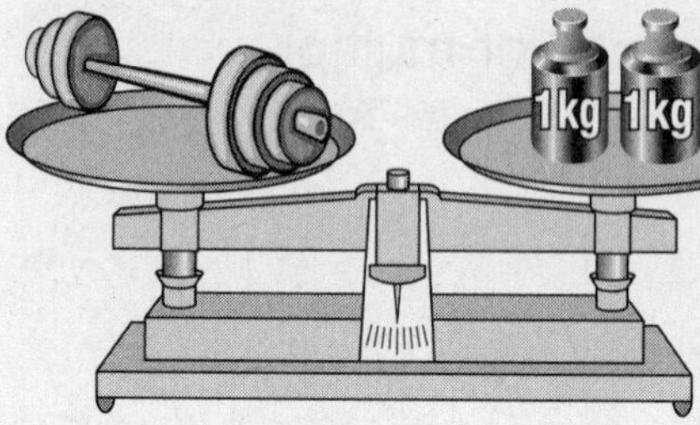

What is the mass of the dumbbell?

- ○ **A.** 2 grams
- ○ **B.** 2 kilograms
- ○ **C.** 20 grams
- ○ **D.** 20 kilograms

2. Which most likely has a mass of 25 kilograms?

○ **A.**

○ **B.**

○ **C.**

○ **D.**

3. Which is the best estimate for the mass of a cotton ball?

- ○ **A.** 1 gram
- ○ **B.** 10 grams
- ○ **C.** 1 kilogram
- ○ **D.** 10 kilograms

4. A chair has a mass of 12 kilograms. A table is 24 kilograms heavier than the chair. What is the mass, in kilograms, of the table?

- ○ **A.** 48 kilograms
- ○ **B.** 36 kilograms
- ○ **C.** 24 kilograms
- ○ **D.** 12 kilograms

5. A box of thumbtacks has a mass of 100 grams. Carolyn bought 2 boxes of thumbtacks. What is the total mass of two boxes of thumbtacks?

- ○ **A.** 20 grams
- ○ **B.** 40 grams
- ○ **C.** 200 grams
- ○ **D.** 400 grams

6. A pencil has a mass of 5 grams. What is the total mass, in grams, of 8 pencils?

- ○ **A.** 10 grams
- ○ **B.** 13 grams
- ○ **C.** 20 grams
- ○ **D.** 40 grams

7. Mr. Marshall baked a large apple pie with a mass of 4 kilograms. He cut the pie into 4 equal pieces. What is the mass, in kilograms, of each piece?

- ○ **A.** 1 kilogram
- ○ **B.** 2 kilograms
- ○ **C.** 4 kilograms
- ○ **D.** 8 kilograms

8. Alex was born at 3 kilograms. On his first birthday, Alex is 8 kilograms.

A. How many kilograms did Alex gain in the first year? Show your work.

____

____

B. On his second birthday, Alex is 4 kilograms heavier than he was at 1 year. What is Alex's mass, in kilograms, on his second birthday? Show your work.

____

____

Common Core State Standard:
3.MD.2

Capacity

Getting the Idea

Capacity is the measure of how much a container can hold.

The table shows two units of capacity in the metric system.

Metric Units of Capacity
1 **liter (L)** = 1,000 **milliliters (mL)**

Use these benchmarks to estimate capacity.

A dropper has a capacity of about 10 milliliters.

A beaker has a capacity of 500 milliliters, or $\frac{1}{2}$ liter.

A sports bottle has a capacity of about 1 liter.

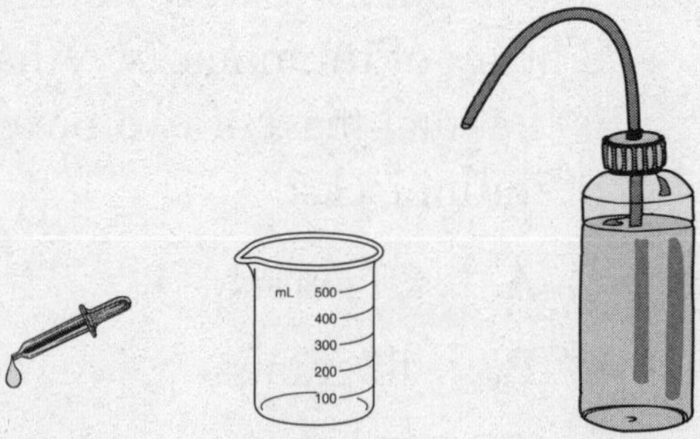

Dropper Beaker Sports Bottle

Below are some containers with different capacities.

4 liters 2 liters 1 liter 250 milliliters

You can measure capacity using milliliters or liters.

Example 1

Jill put water in the beaker.

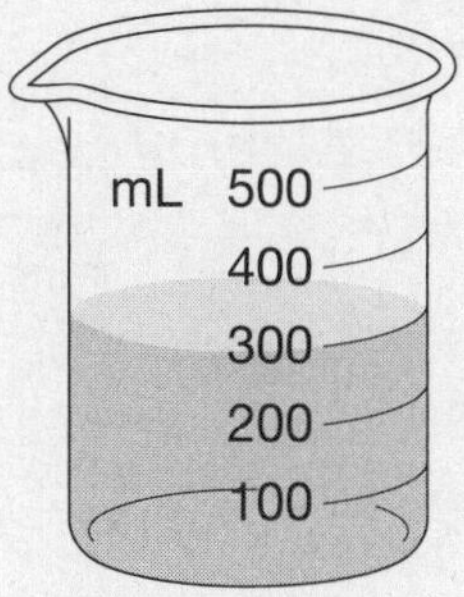

How much water is in the beaker?

Strategy **Find the amount of water in the beaker.**

Step 1 Look at the marks on the beaker.

Each mark is 100 milliliters.

Step 2 Read the mark that the water comes up to.

The water stops at the 300-milliliter mark.

Solution **There are 300 milliliters of water in the beaker.**

Example 2

Which is the better estimate for the capacity of a can of juice?

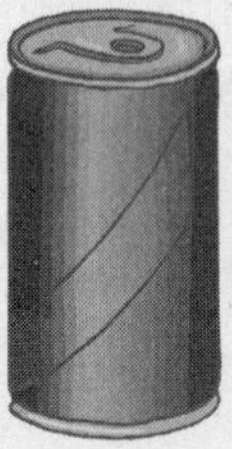

300 liters 300 milliliters

Strategy **Look at the units in the choices.**
Compare the units to the juice can.

A can of juice holds less than 1 liter.

So, the units must be milliliters.

Solution **The capacity of a can of juice is about 300 milliliters.**

Example 3

Dina put liquid in the beakers.

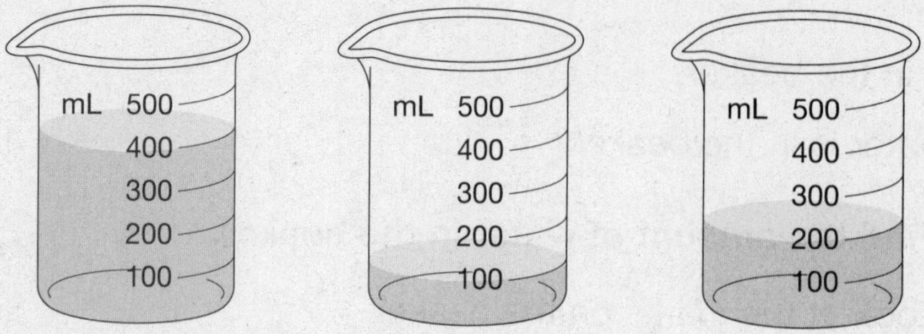

How much liquid did Dina put in the beakers in all?

Strategy **Look for key words to decide which operation to use.**

Step 1 Decide which operation to use.

The phrase "in all" means add.

Write an addition sentence.

Use $\square$ for the sum.

$400 + 100 + 200 = \square$

Step 2 Add.

$$\begin{array}{r} 400 \\ 100 \\ +\ 200 \\ \hline 700 \end{array}$$

Solution **There are 700 milliliters of liquid in all.**

Example 4

Erica bought 8 bottles of lemonade for a party.
Each bottle contains 2 liters (L) of lemonade.

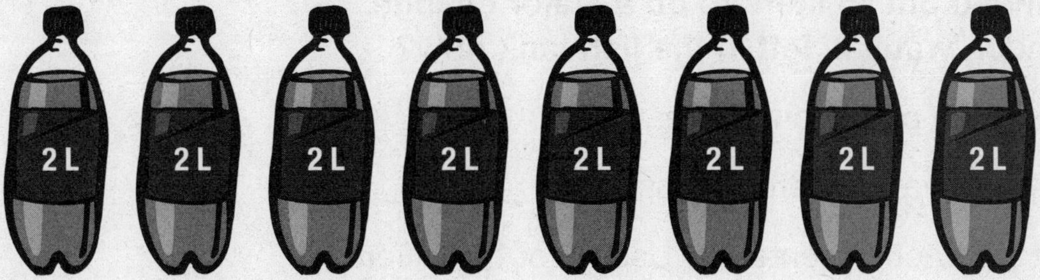

How many liters of lemonade did Erica buy in all?

Strategy **Look for key words to decide which operation to use.**

Step 1 Decide which operation to use.

"How many liters in all" means to add or multiply.

Multiplying is faster.

There are 8 bottles, and each bottle is 2 liters.

$8 \times 2 = \square$

Step 2 Multiply.

$8 \times 2 = 16$

Solution **Erica bought 16 liters of lemonade in all.**

Coached Example

Jared's fish tank has 258 liters of water.
He pumped out 65 liters to do a water change.
How much water is left in the fish tank now?

Decide which operation to use.

"How much water is left" means ______________________.

Write a subtraction sentence. Use □ for the difference.

__________ – __________ = □

Find the difference.

Use addition to check your answer.

The fish tank has __________ liters of water left.

Lesson Practice

Choose the correct answer.

1. Marco filled this beaker with a liquid.

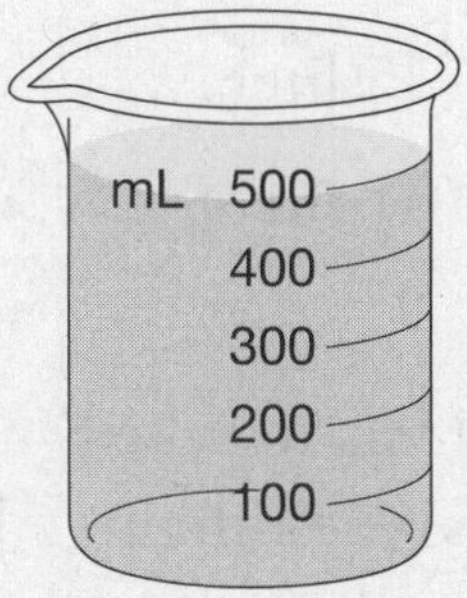

About how much liquid is in the beaker?

- ○ **A.** 100 milliliters
- ○ **B.** 200 milliliters
- ○ **C.** 400 milliliters
- ● **D.** 500 milliliters

2. Harold poured some milk into a bowl with cereal. Which measure is most likely the amount of milk that Harold poured into the bowl?

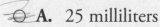

- ○ **A.** 25 milliliters
- ○ **B.** 25 liters
- ○ **C.** 250 milliliters
- ○ **D.** 250 liters

3. Which container could have a capacity of 10 liters?

○ **A.**

● **B.**

○ **C.**

○ **D.**

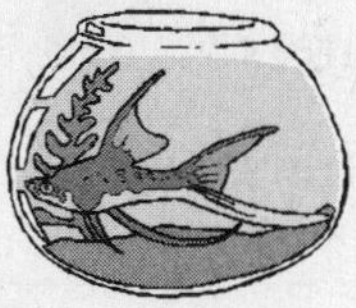

4. A bathtub has 145 liters of water in it. Cindy drained 38 liters of water. About how much water is in the bathtub now?

- ○ **A.** 107 liters
- ○ **B.** 113 liters
- ○ **C.** 117 liters
- ○ **D.** 183 liters

5. Before lunch, Mr. Jones filled the kiddy pool with 350 liters of water. After lunch, he filled the pool with 585 liters of water. About how much water did Mr. Jones fill in the pool?

○ **A.** 235 liters

○ **B.** 700 liters

○ **C.** 835 liters

○ **D.** 935 liters

6. Chuck drank a total of 7 liters of water in 7 days. If he drank the same amount of water each day, about how much water did he drink each day?

○ **A.** 1 liter

○ **B.** 7 liters

○ **C.** 14 liters

○ **D.** 49 liters

7. Shannon used 67 liters of water in the shower. Nathan used 58 liters of water in the shower. About how much water did they use in all?

○ **A.** 115 liters

○ **B.** 125 liters

○ **C.** 155 liters

○ **D.** 161 liters

8. Michael has to take 8 milliliters of medicine each night. About how much medicine does Michael take in all for 7 nights?

○ **A.** 1 milliliter

○ **B.** 15 milliliters

○ **C.** 56 milliliters

○ **D.** 78 milliliters

9. Jasmine bought 10 bottles of soda for a barbeque at the park. Each bottle has 2 liters of soda.

A. How many liters of soda did Jasmine buy in all? Show your work.

____

B. Jasmine will put the same amount of soda on each table. There are 5 tables. How many liters of soda does each table get? Show your work.

____

Common Core State Standard:
3.MD.8

Perimeter

Getting the Idea

Perimeter is the distance around a figure.

Some units for measuring perimeter are inches, centimeters, feet, and yards.

You can add the lengths of the sides to find the perimeter of a figure.

Example 1

What is the perimeter of this triangle?

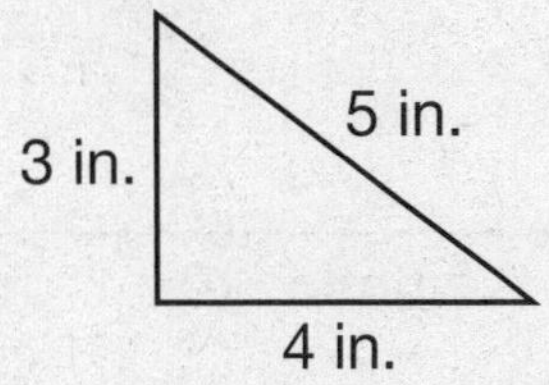

Strategy **Use addition.**

Step 1 Add the lengths of the sides.

$3 + 4 + 5 = 12$

Step 2 Look at the units labeled on the triangle.

The units are inches.

Solution **The perimeter of the triangle is 12 inches.**

Some problems may not give the measurements of a figure.

You will need to measure the lengths of the sides.

Then add the lengths to find the perimeter.

Example 2

Laida used this figure to trace rectangles on a poster.

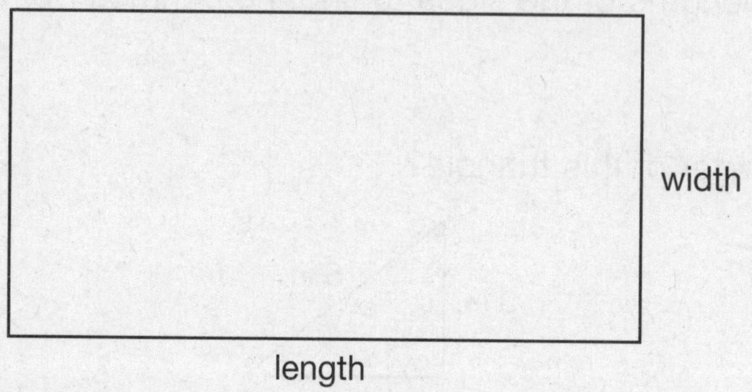

What is the perimeter of this rectangle, in centimeters?

Strategy **Use a centimeter ruler to measure each side. Then add the measurements.**

Step 1 Measure the length of the rectangle.

The length is 8 centimeters.

Two sides of the rectangle are 8 centimeters long.

Step 2 Measure the width of the rectangle.

The width is 4 centimeters.

Two sides of the rectangle are 4 centimeters long.

Step 3 Add the measurements.

$8 + 8 + 4 + 4 = 24$

Solution **The perimeter of the rectangle is 24 centimeters.**

You can also use multiplication to find the perimeter of a figure.

A square is a figure with 4 equal sides. To find the perimeter of a square, you can multiply 4 by the length of one side.

Perimeter of a square $= 4 \times$ length of side

Example 3

What is the perimeter of this square?

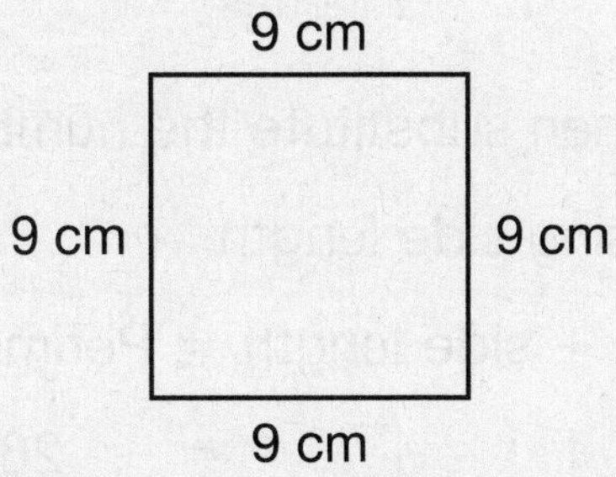

Strategy **Use multiplication.**

Step 1 Multiply 4 by the length of one side.

$4 \times 9 = 36$

Step 2 Look at the units labeled on the square.

The units are centimeters.

Solution **The perimeter of the square is 36 centimeters.**

You can write a number sentence to help you find the missing length in a perimeter problem.

The triangle below has a perimeter of 26 inches. Find the missing side length.

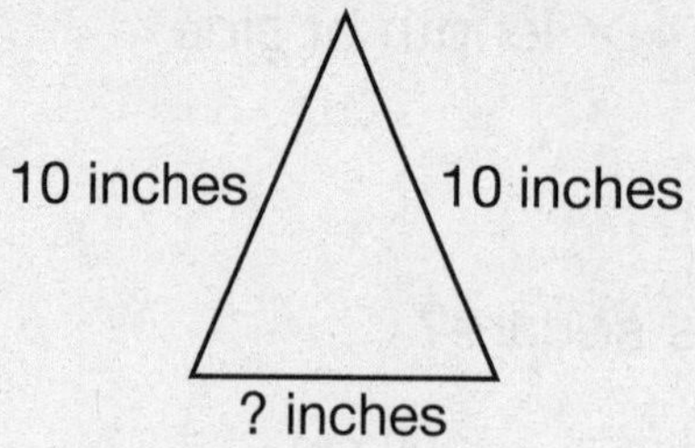

Write a number sentence. Then substitute the numbers you know.

Use $\square$ to represent the missing side length.

side length + side length + side length = Perimeter

$10 + 10 + \square = 26$

Find the missing side length.

$10 + 10 + \square = 26$	Add the two side lengths.
$20 + \square = 26$	Think: $20 + ? = 26$
$20 + 6 = 26$	

The missing side length is 6 inches.

Example 4

This hexagon has 6 equal sides. It has a perimeter of 42 meters.

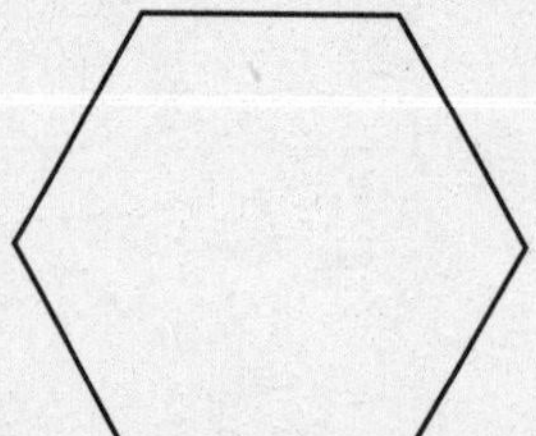

What is the side length of the hexagon?

Strategy **Write a number sentence. Then substitute the numbers you know.**

Step 1 Write a number sentence.

Use □ to represent the missing side length.

The hexagon has 6 equal sides.

So, the perimeter is 6 × the length of one side.

Perimeter = 6 × □

Step 2 Find the side length.

Substitute the numbers you know.

42 = 6 × □

Think: 6 × ? = 42

6 × 7 = 42

The side length is 7 meters.

Solution **The side length of the hexagon is 7 meters.**

Coached Example

What is the perimeter of this rectangle?

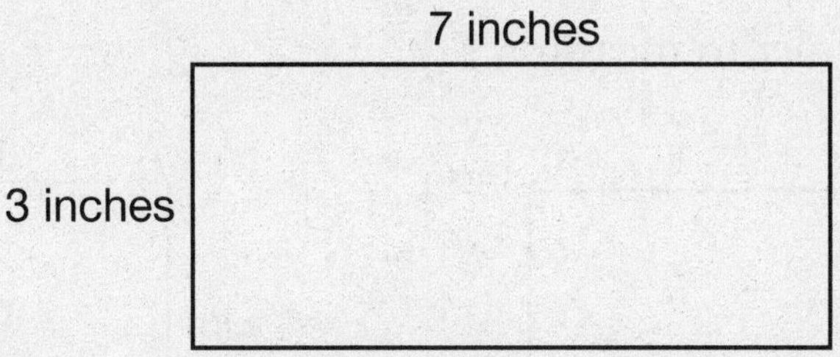

The rectangle has ________ sides.

Two sides of the rectangle are ________ inches long.

The other two sides of the rectangle are ________ inches long.

Add the measurements.

________ + ________ + ________ + ________ = ________

The perimeter of the rectangle is ________ inches.

Lesson Practice

Choose the correct answer.

1. What is the perimeter of this triangle?

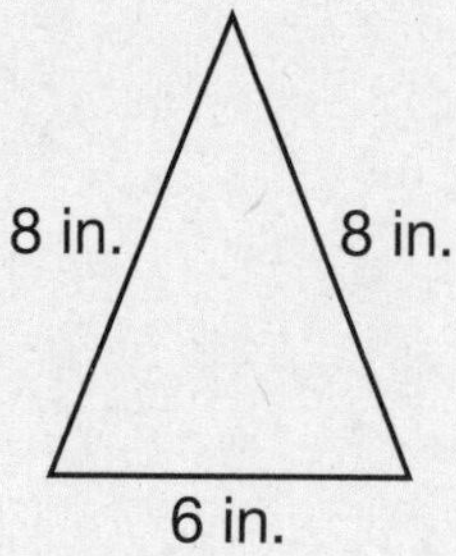

○ **A.** 18 inches
○ **B.** 20 inches
○ **C.** 22 inches
○ **D.** 24 inches

2. Use a centimeter ruler to measure the side lengths.

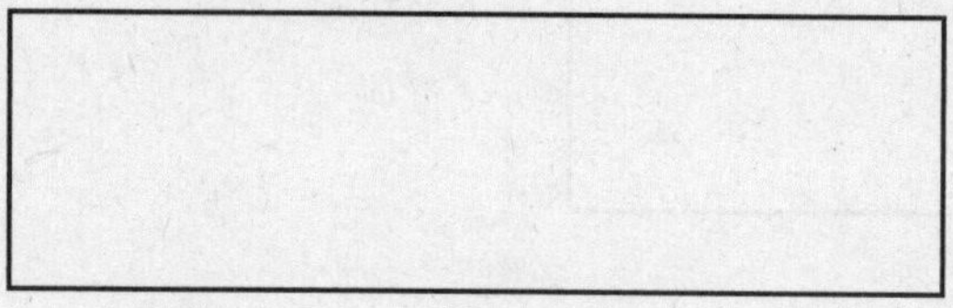

What is the perimeter of this rectangle?

○ **A.** 18 centimeters
○ **B.** 14 centimeters
○ **C.** 9 centimeters
○ **D.** 7 centimeters

3. Which rectangle has a different perimeter than the others?

○ **A.**

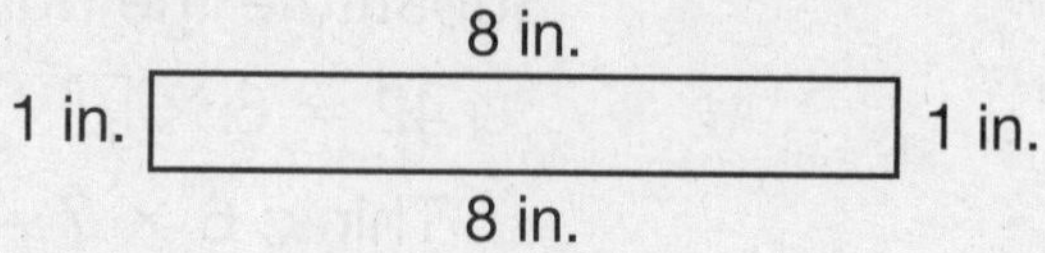

○ **B.**

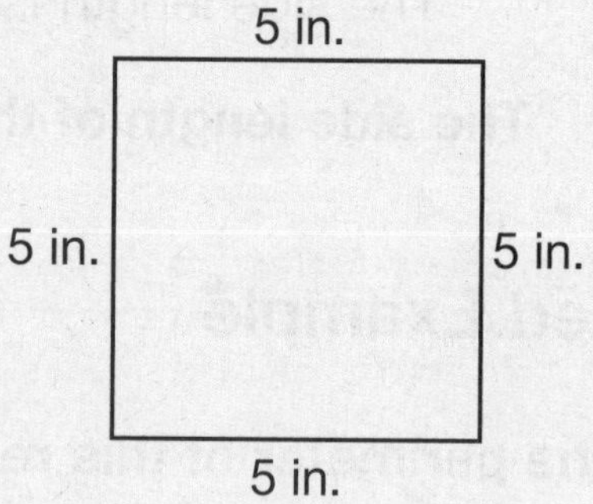

○ **C.**

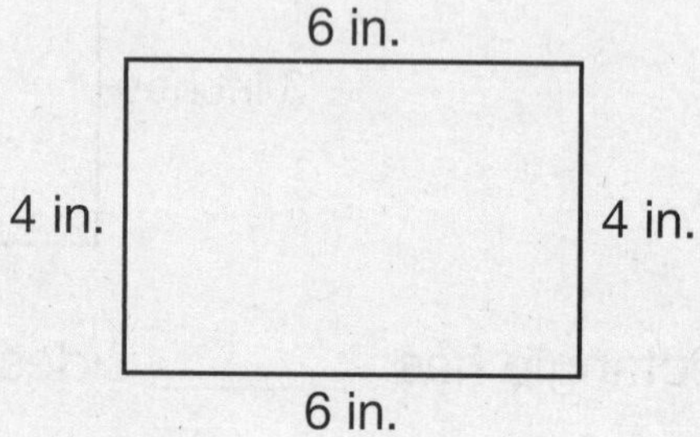

○ **D.**

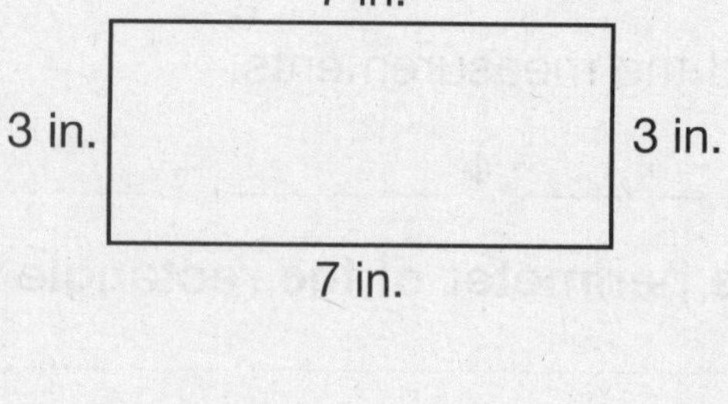

4. The pentagon below has sides of equal length.

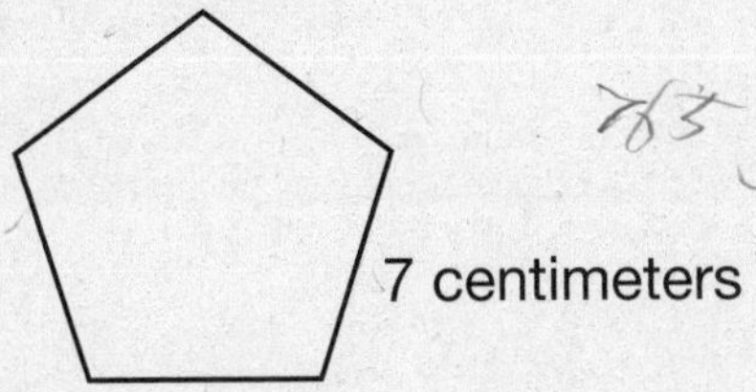

What is the perimeter of the pentagon?

- ○ **A.** 14 centimeters
- ○ **B.** 21 centimeters
- ○ **C.** 35 centimeters
- ○ **D.** 42 centimeters

5. A playground is shaped like a rectangle. It has a length of 9 yards and a width of 8 yards. What is the perimeter of the playground?

- ○ **A.** 18 yards
- ○ **B.** 32 yards
- ○ **C.** 34 yards
- ○ **D.** 36 yards

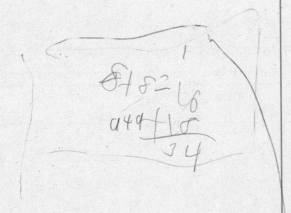

6. The perimeter of this rectangle is 18 meters.

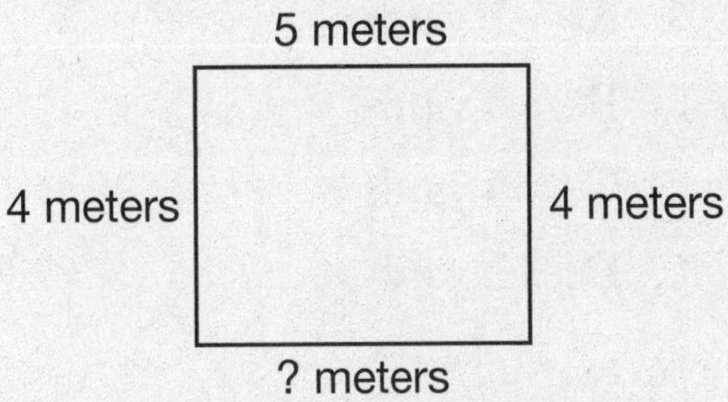

Which number sentence can be used to find the missing side length?

- ○ **A.** 4 + 5 + □ = 18
- ○ **B.** 4 + 4 + 4 + □ = 18
- ○ **C.** 5 + 5 + 5 + □ = 18
- ○ **D.** 4 + 5 + 4 + □ = 18

7. This figure has a perimeter of 24 feet.

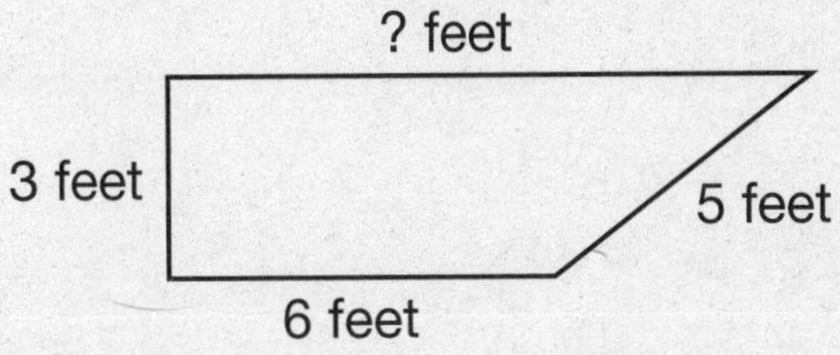

What is the missing side length?

- ○ **A.** 6 feet
- ○ **B.** 10 feet
- ○ **C.** 11 feet
- ○ **D.** 12 feet

8. A square rug has a perimeter of 8 yards. What is the length of one side of the rug?

- ○ **A.** 2 yards
- ○ **B.** 3 yards
- ○ **C.** 4 yards
- ● **D.** 12 yards

9. Adam has these two posters. Poster A is a rectangle. Poster B is a square.

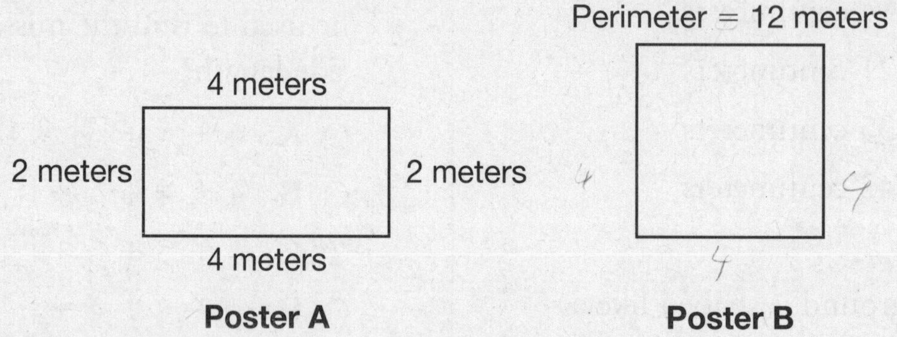

A. What is the perimeter of poster A? Show your work.

4x4=16

2x2=4

16
+4
20

20 meters

B. What is the length of each side of poster B? Explain your answer.

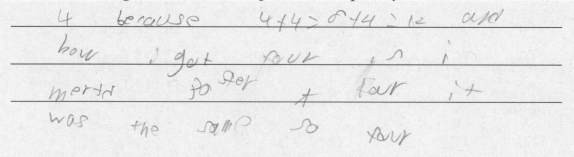

4 because 4+4=8+4=12 and how i got four is i mertd to ster at four it was the same so four

Common Core State Standards:
3.MD.5.a, 3.MD.5.b, 3.MD.6

Understand Area

Getting the Idea

Area is the number of **square units** needed to cover a figure.

A square with a side length of 1 unit is a unit square.

For example, 1 square inch is a square with side lengths of 1 inch.

1 square centimeter is a square with side lengths of 1 centimeter.

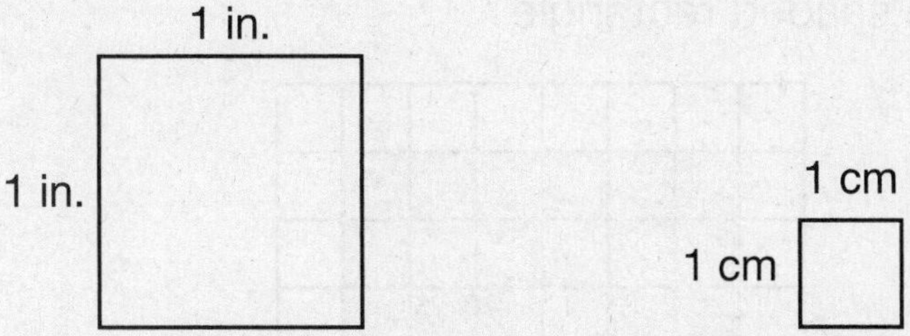

1 square inch
1 in.²

1 square centimeter
1 cm²

Other examples of square units are square feet and square meters.

To find the area, you can count the number of square units that cover the figure.

Example 1

Arthur used square tiles to make a rectangle. Each  is a square with side lengths of 1 centimeter.

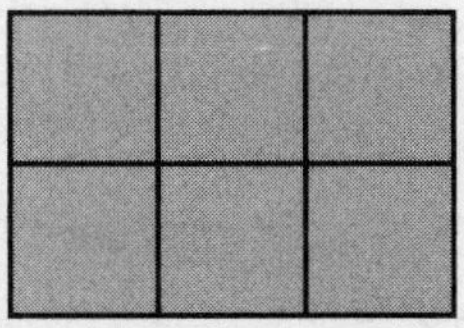

What is the area of the rectangle?

Strategy **Count the number of square units that make up the rectangle.**

Step 1 The rectangle is made up of 6 square tiles.

Step 2 Find the area.

Each ■ is 1 square centimeter.

So, 6 ■ are 6 square centimeters.

Solution **The area of the rectangle is 6 square centimeters or 6 cm^2.**

Example 2

What is the area of the shaded rectangle?

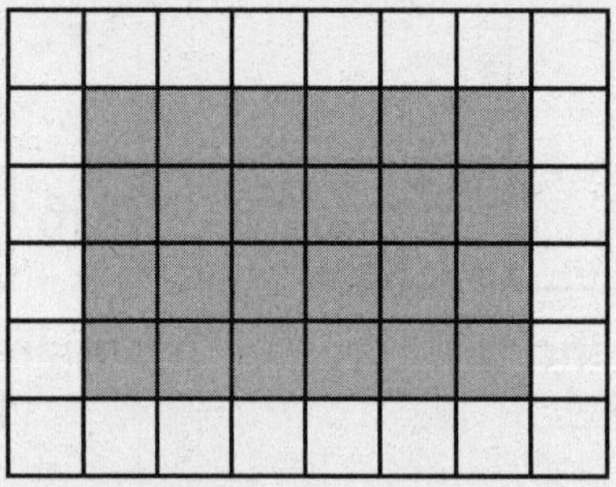

Key: ■ = 1 square foot

Strategy **Count the number of shaded squares in each row. Then add.**

Step 1 Count the number of rows and the shaded squares in each row.

There are 4 rows of shaded squares.

Each row has 6 shaded squares.

Step 2 Use repeated addition.

Add 6 four times.

$6 + 6 + 6 + 6 = 24$

Step 3 Write the units.

Each ■ = 1 square foot.

24 ■ equal 24 square feet.

Solution **The area of the shaded rectangle is 24 square feet or 24 ft^2.**

Example 3

How much greater is the area of the shaded rectangle in Example 2 than the area of the shaded rectangle below? The area of the shaded rectangle in Example 2 is 24 square feet.

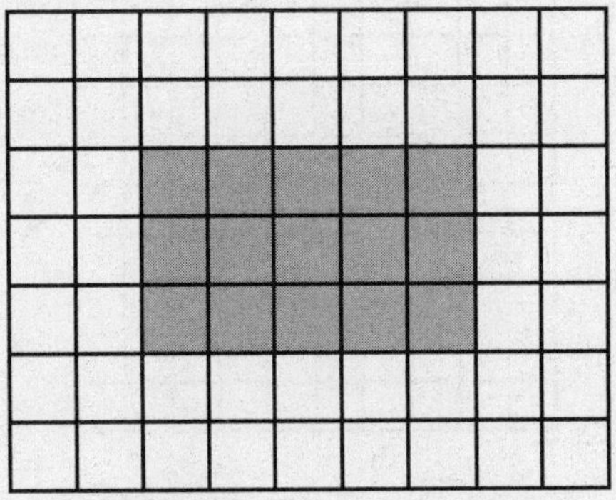

Key:  = 1 square foot

Strategy **First, find the area of this rectangle. Then subtract from the area of the rectangle in Example 2.**

Step 1 Count the number of rows and the shaded squares in each row.

There are 3 rows of 5 shaded squares.

Use repeated addition.

Add 5 three times.

$5 + 5 + 5 = 15$

Step 2 Find the area and label the units.

Each  = 1 square foot.

15  equal 15 square feet.

Step 3 Subtract 15 square feet from the area of the rectangle in Example 2.

The area of the rectangle in Example 2 is 24 square feet.

$24 - 15 = 9$ square feet

Solution **The area of the rectangle in Example 2 is 9 square feet greater than the area of the rectangle above.**

Coached Example

What is the area of the shaded figure?

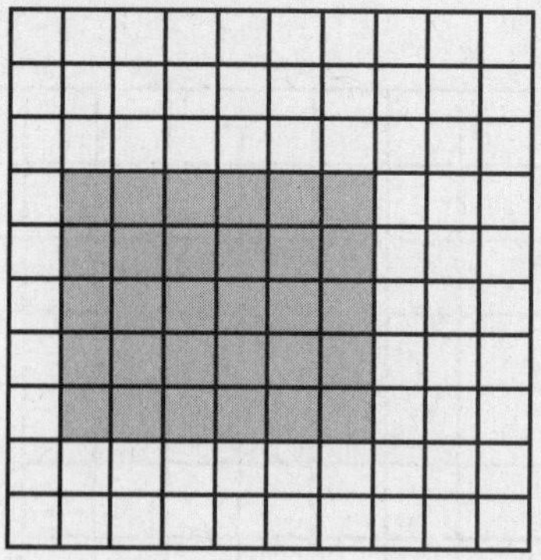

Key: ▢ = 1 square meter

Count the number of rows and the shaded squares in each row.

There are ________ rows of shaded squares.

Each row has ________ shaded squares.

Use repeated addition.

Add ________ five times.

________ + ________ + ________ + ________ + ________ = ________

What are the units for the area? ____________________

The area of the shaded figure is ________ square meters or ________ m^2.

Lesson Practice

Choose the correct answer.

1. Each ■ equals 1 square inch.

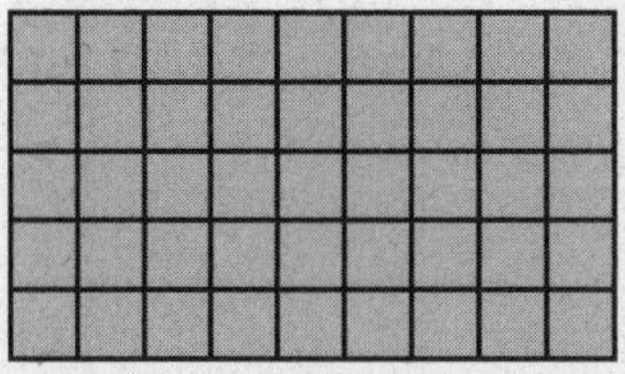

What is the area of the rectangle?

- ○ **A.** 45 square inches
- ○ **B.** 40 square inches
- ○ **C.** 36 square inches
- ○ **D.** 28 square inches

2. Chloe's playroom is a square with side lengths of 10 feet.

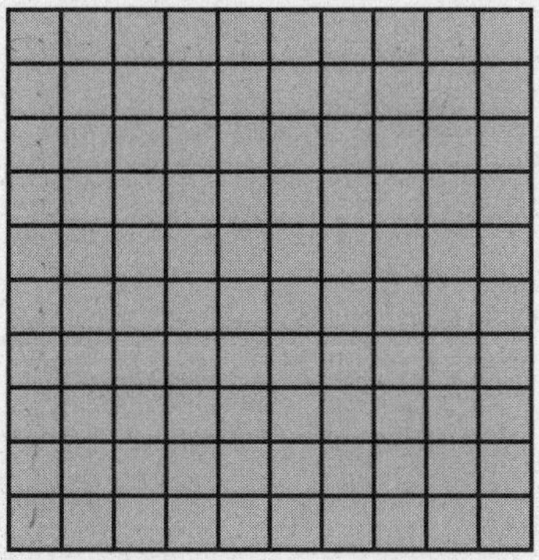

Key: ■ = 1 square foot

What is the area of the playroom?

- ○ **A.** 10 square feet
- ○ **B.** 20 square feet
- ○ **C.** 40 square feet
- ○ **D.** 100 square feet

3. What is the area of the shaded figure?

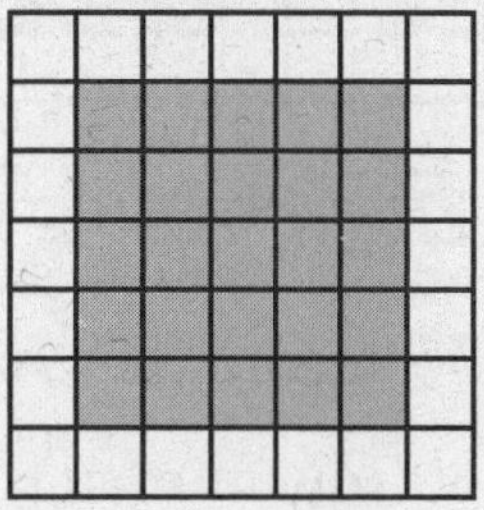

Key: ■ = 1 square centimeter

- ○ **A.** 5 square centimeters
- ○ **B.** 25 square centimeters
- ○ **C.** 35 square centimeters
- ○ **D.** 49 square centimeters

4. What is the area of the shaded rectangle?

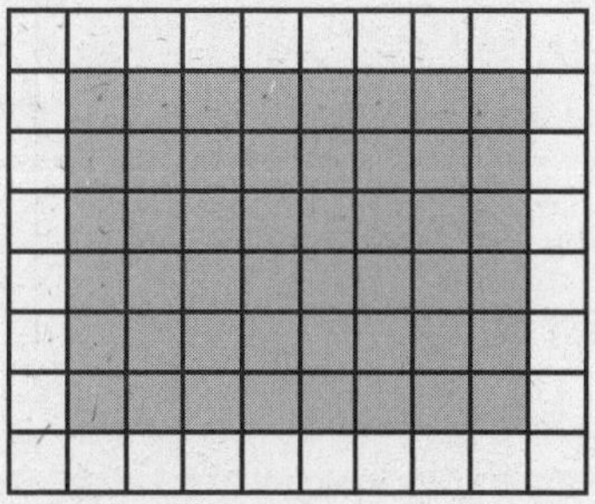

Key: ■ = 1 square foot

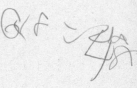

- ○ **A.** 28 square feet
- ○ **B.** 32 square feet
- ○ **C.** 40 square feet
- ○ **D.** 48 square feet

5. What is the area of the shaded figure?

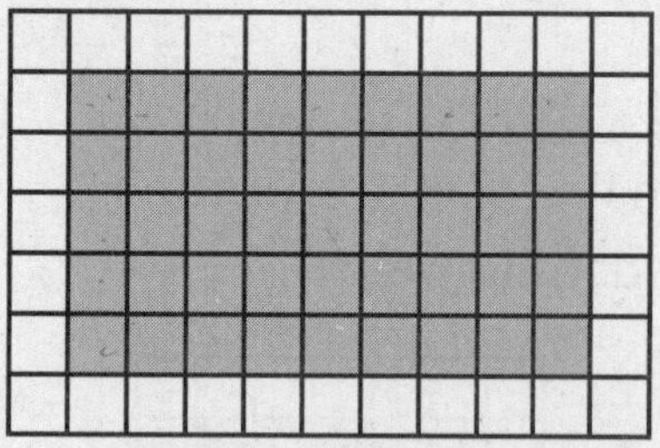

Key: ▢ = 1 square meter

- A. 36 square meters
- B. 40 square meters
- C. 45 square meters
- D. 77 square meters

5x9=45

6. What is the area of the shaded figure?

9x9=81

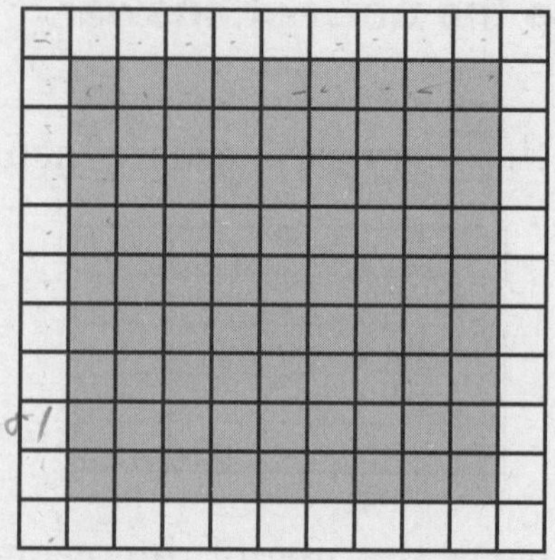

Key: ▢ = 1 square centimeter

- A. 9 square centimeters
- B. 18 square centimeters
- C. 27 square centimeters
- D. 81 square centimeters

7. Thad drew a rectangle on the grid.

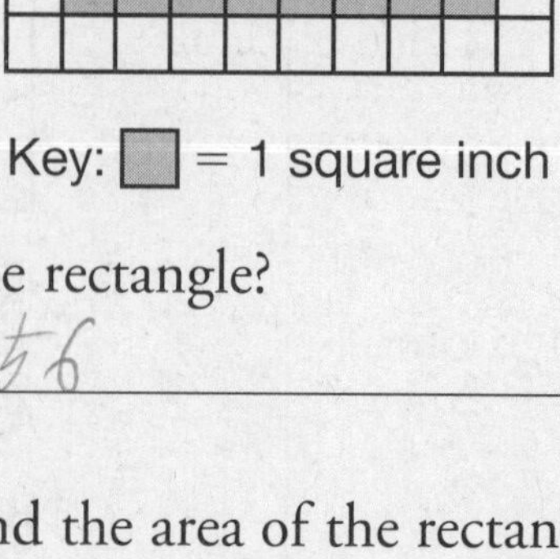

Key: ▢ = 1 square inch

A. What is the area of the rectangle?

7x8=56

B. Explain how you found the area of the rectangle in Part A.

in the black square has 7 and 8 7 from side 8 from up

7x8=56

Common Core State Standards:
3.MD.7.a, 3.MD.7.b, 3.MD.7.c, 3.MD.7.d

Area of Rectangles

Getting the Idea

The area of a rectangle is the number of square units that cover the rectangle.

The rectangle below has 3 rows of squares.

Each row has 5 square inches.

The area is 5 + 5 + 5 = 15 square inches.

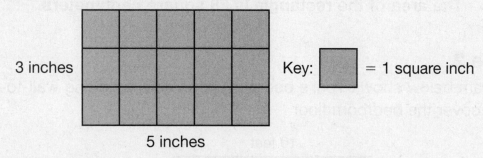

3 inches

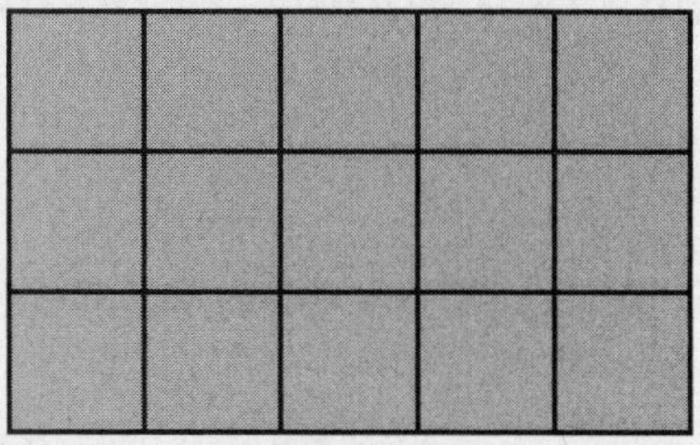

5 inches

Key: ▢ = 1 square inch

You can also multiply the length and the width to find the area.

5 inches × 3 inches = 15 square inches

Example 1

What is the area of the shaded rectangle?

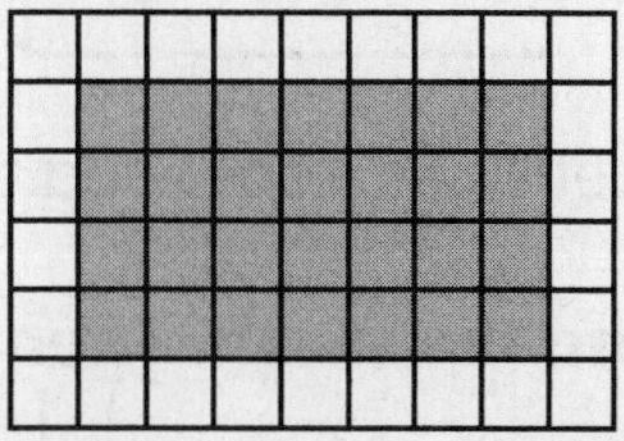

Key: ▢ = 1 square centimeter

Strategy **Use repeated addition or multiplication.**

Step 1 Count the number of rows and the number in each row.

There are 4 rows of 7 squares.

Step 2 Use repeated addition or multiplication.

Add 7 four times.

$7 + 7 + 7 + 7 = 28$

$4 \times 7 = 28$

Step 3 Use the scale key to find what each ■ represents.

Each ■ equals 1 square centimeter.

So, 28 ■ = 28 square centimeters.

Solution **The area of the rectangle is 28 square centimeters.**

Example 2

The diagram below shows Toni's bedroom floor. She is getting wall-to-wall carpet to cover the bedroom floor.

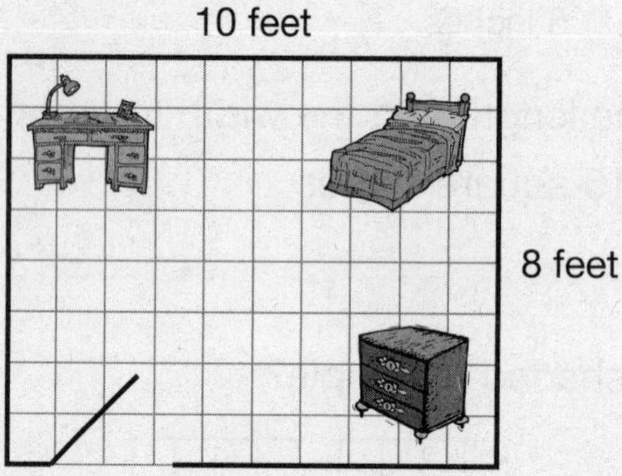

How many square feet of carpet does Toni need to cover her bedroom floor?

Strategy **Find the length and the width. Then multiply.**

Step 1 Find the length.

The length is 10 feet.

Step 2 Find the width.

The width is 8 feet.

Step 3 Multiply the length times the width.

10 feet $\times$ 8 feet = 80 square feet

Solution **Toni needs 80 square feet of carpet for her bedroom floor.**

You can use the **distributive property** to find the area of a rectangle. Remember, the distributive property says that multiplying a sum by a factor is the same as multiplying each addend by the factor and adding the products.

For example, find the area of a rectangle with width 3 inches and length 12 inches.

You can break the rectangle into 2 rectangles and use the distributive property.

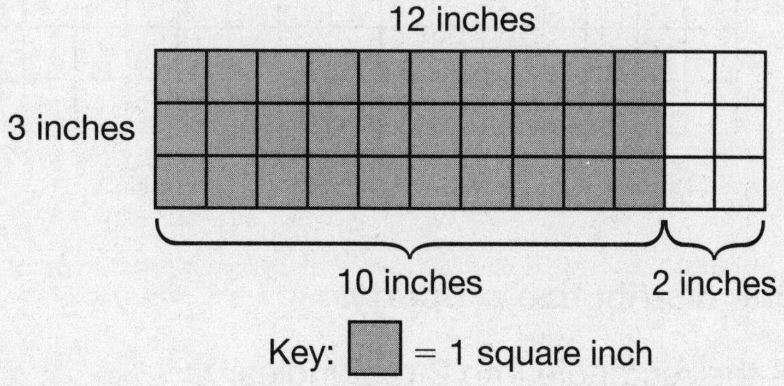

Rename one of the factors.	$3 \times (10 + 2)$
Multiply the other factor by each addend.	$(3 \times 10) + (3 \times 2)$
Add the products.	$30 + 6$

The area of the rectangle is 36 square inches.

Example 3

What is the area of a rectangle with width 5 centimeters and length 15 centimeters?

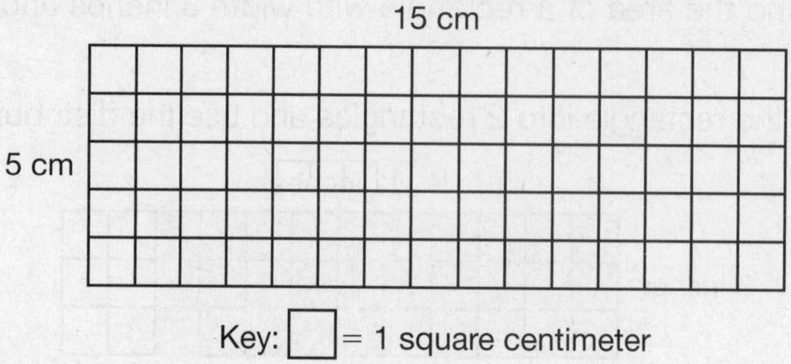

Strategy **Use the distributive property.**

Step 1 Break the rectangle into 2 rectangles.

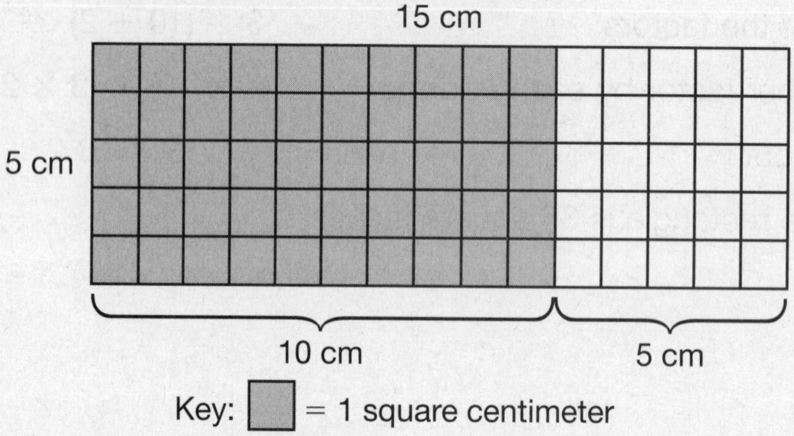

$5 \times 15 = 5 \times (10 + 5)$

Step 2 Use the distributive property.

$5 \times (10 + 5)$

$(5 \times 10) + (5 \times 5)$

$50 + 25 = 75$

So, the area of the rectangle is 75 square centimeters.

Solution **The area of a rectangle with width 5 centimeters and length 15 centimeters is 75 square centimeters.**

Example 4

What is the area of the figure below?

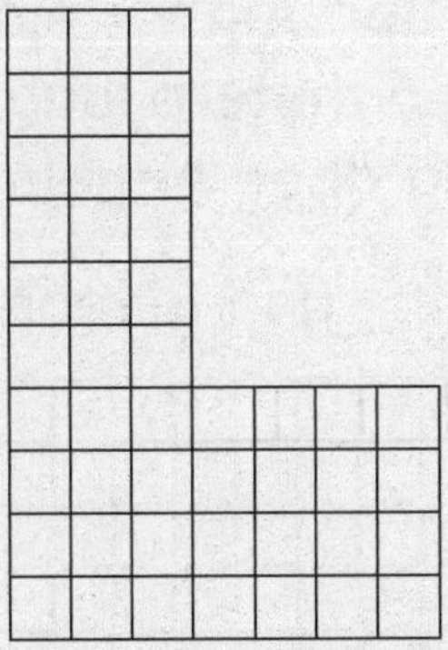

Key: □ = 1 square inch

Strategy **Break the figure into 2 rectangles. Find the area of each rectangle. Add to find the total area.**

Step 1 Break the figure into rectangle *A* and rectangle *B*.

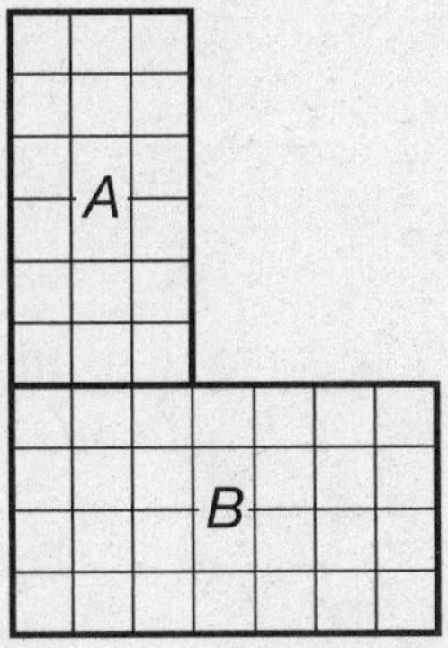

Key: □ = 1 square inch

Step 2 Multiply to find the area of each rectangle.

Rectangle A has a length of 3 inches and a width of 6 inches.

3 inches × 6 inches = 18 square inches

Rectangle B has a length of 7 inches and a width of 4 inches.

7 inches × 4 inches = 28 square inches

Step 3 Add to find the total area.

18 square inches + 28 square inches = 46 square inches

Solution **The area of the figure is 46 square inches.**

Coached Example

Angelo wants to buy linoleum flooring for his kitchen. The kitchen is shaped like a rectangle that is 5 meters long and 4 meters wide. How many square meters of flooring does Angelo need?

Draw and shade a rectangle on the grid to represent the kitchen floor.

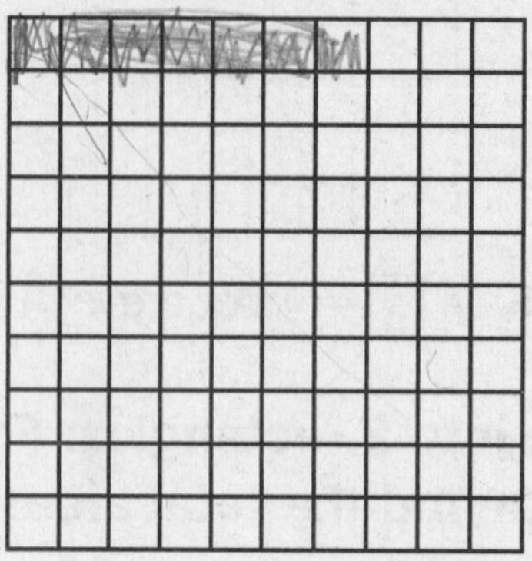

Key: □ = 1 square meter

How many squares did you shade? __________

Use multiplication to find the area.

You need to multiply the __________________ times the __________________.

__________ × __________ = __________

The units for the area of the floor are ________________ ________________.

Angelo needs __________ square meters of flooring.

Lesson Practice

Choose the correct answer.

1. Kenny made a drawing of the garden he is planning.

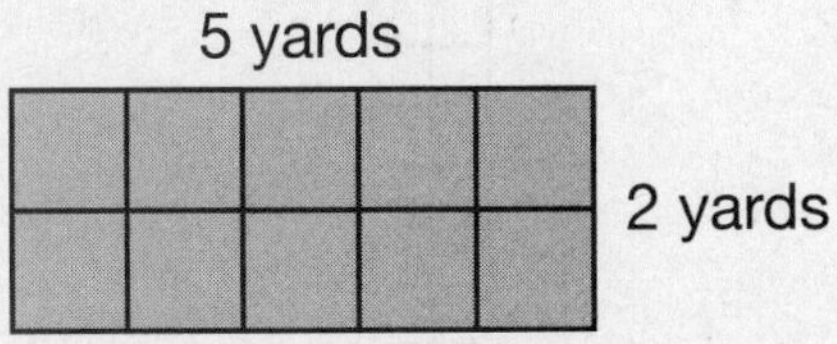

What is the area of Kenny's garden?

- ○ **A.** 3 square yards
- ○ **B.** 7 square yards
- ○ **C.** 10 square yards
- ○ **D.** 14 square yards

2. Ratisha marked an area for a handball court in her driveway.

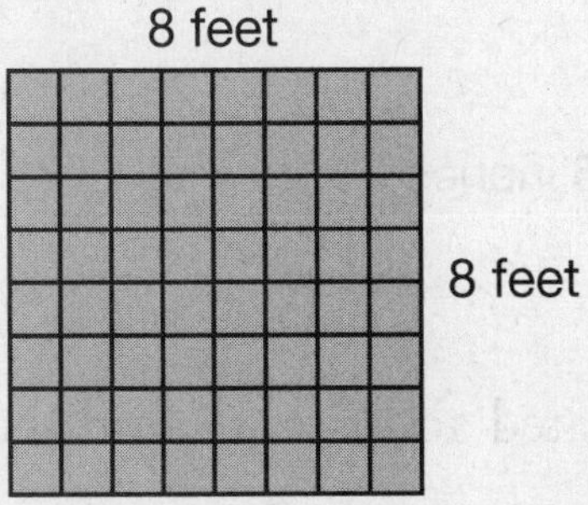

What is the area of Ratisha's handball court?

- ○ **A.** 32 square feet
- ○ **B.** 64 square feet
- ○ **C.** 88 square feet
- ○ **D.** 100 square feet

3. Lorenzo has a photo frame with an area of 80 square inches. The width of the frame is 10 inches.

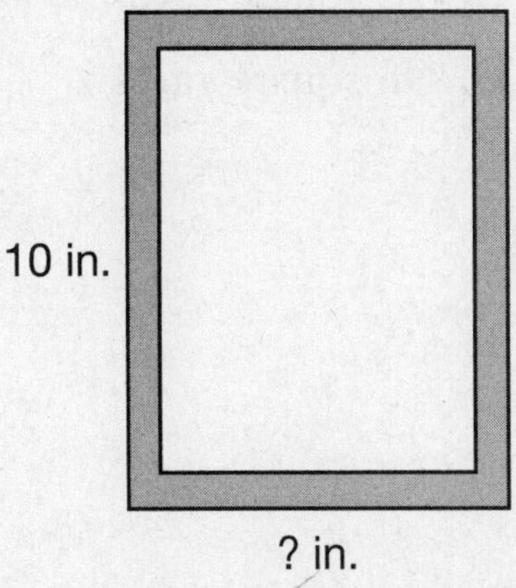

What is the length of the frame?

- ○ **A.** 4 inches
- ○ **B.** 8 inches
- ○ **C.** 30 inches
- ○ **D.** 40 inches

4. A room is 6 yards long and 5 yards wide. What is the area of the room?

- ○ **A.** 22 square yards
- ○ **B.** 26 square yards
- ○ **C.** 30 square yards
- ○ **D.** 32 square yards

5. A rectangular blacktop has a length of 9 yards and a width of 8 yards. What is the area of the blacktop?

- ○ **A.** 17 square yards
- ○ **B.** 34 square yards
- ○ **C.** 72 square yards
- ○ **D.** 98 square yards

6. The ninth hole at Ralph's Mini Golf is shaped like the figure below.

Key: □ = 1 square foot

What is the area of the figure?

- ○ **A.** 34 square feet
- ○ **B.** 36 square feet
- ○ **C.** 40 square feet
- ○ **D.** 80 square feet

7. Carlotta built a flower box for her window.

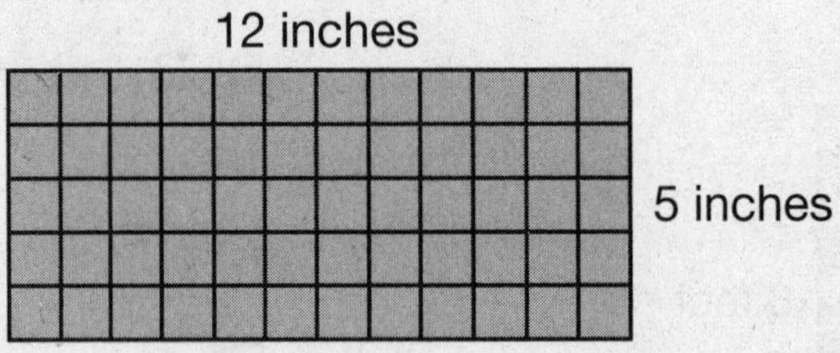

A. What is the area of the flower box? Use repeated addition to find the area.

____

B. What is the area of the flower box? Use the distributive property to find the area.

____

____

Compare Perimeter and Area

Common Core State Standard:
3.MD.8

Getting the Idea

Rectangles can have the same perimeter, but different areas.

They can also have the same areas, but different perimeters.

These rectangles all have an area of 12 square units, but they have different perimeters.

Perimeter = 26 units

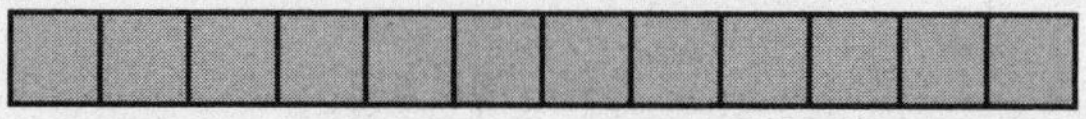

Perimeter = 16 units

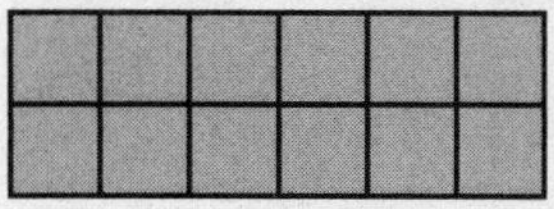

Perimeter = 14 units

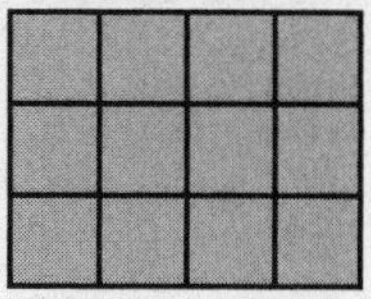

Example 1

What are the perimeters and areas of these two shapes?

Shape A

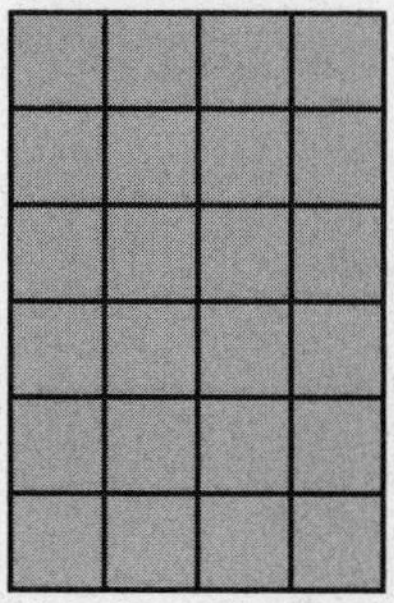

Shape B

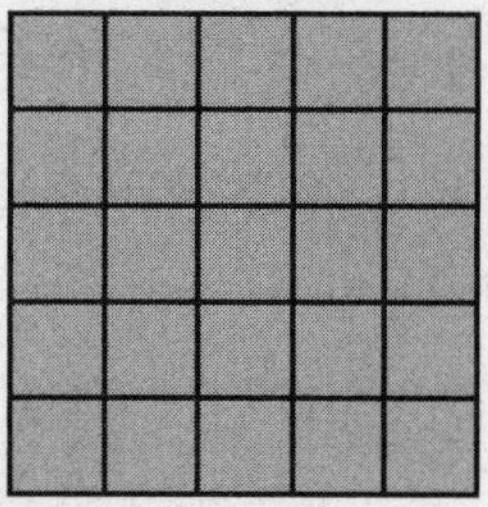

Strategy **Add the side lengths to find the perimeter. Multiply the side lengths to find the area.**

Step 1 Find the perimeter and area of Shape A.

It has a length of 4 units. It has a width of 6 units.

The perimeter is $4 + 6 + 4 + 6 = 20$ units.

The area is $4 \times 6 = 24$ square units.

Step 2 Find the perimeter and area of Shape B.

It has a length of 5 units. It has a width of 5 units.

The perimeter is $5 + 5 + 5 + 5 = 20$ units.

The area is $5 \times 5 = 25$ square units.

Solution **Both shapes have a perimeter of 20 units. Shape A has an area of 24 square units. Shape B has an area of 25 square units.**

Example 2

Janelle labeled the rectangle below.

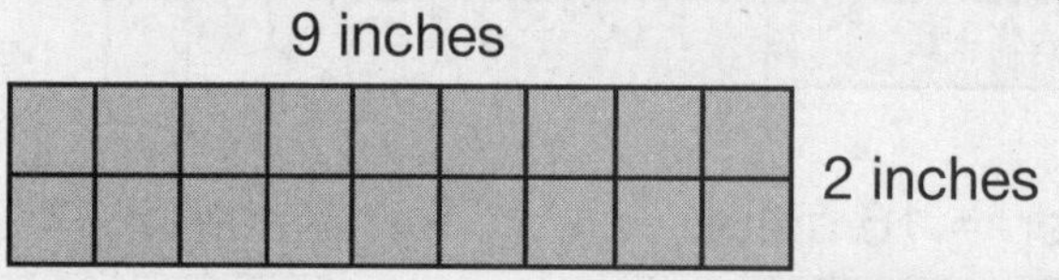

Make a rectangle with the same area as Janelle's rectangle, but with a different perimeter.

Strategy **Use square tiles.**

Step 1 Find the perimeter and area of Janelle's rectangle.

It has a length of 9 inches and a width of 2 inches.

It has a perimeter of $9 + 2 + 9 + 2 = 22$ inches.

It has an area of 9 inches $\times$ 2 inches $=$ 18 square inches.

Step 2 Use 18 square tiles to represent 18 square inches.

Make a different rectangle with 18 square tiles.

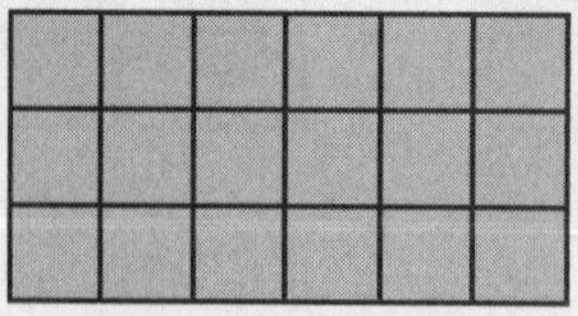

Step 3 Check that the perimeter is different.

The shape has a length of 6 inches and a width of 3 inches.

It has a perimeter of $6 + 3 + 6 + 3 = 18$ inches.

Solution **A shape with a length of 6 inches and a width of 3 inches has the same area as, but a different perimeter than, Janelle's rectangle.**

Coached Example

Mark made a poster for the Science Fair. It was shaped like a rectangle and was 5 feet long and 2 feet wide.

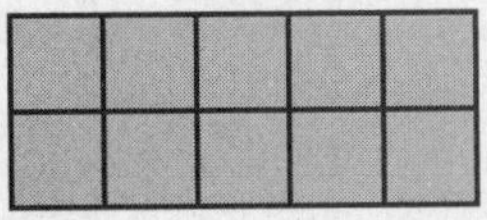

What are the measurements of a rectangle with the same perimeter but different area than Mark's poster?

Find the perimeter and area of Mark's poster.

It has a length of ________ feet and a width of ________ feet.

Find the perimeter.

________ + ________ + ________ + ________ = ________ feet

Find the area.

________ feet × ________ feet = ________ square feet

Make a rectangle with the same perimeter but with a different area.

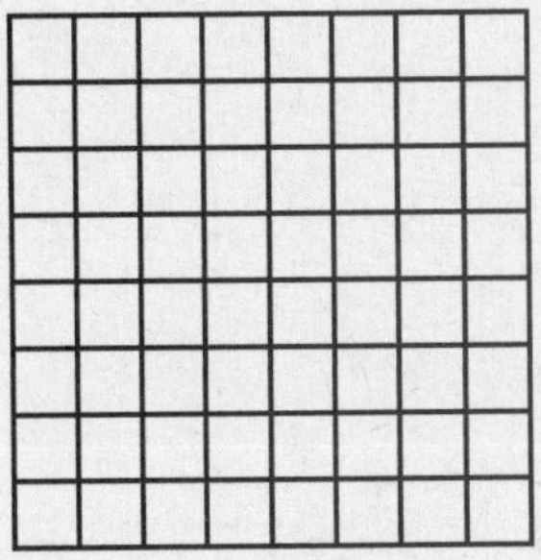

Key: □ = 1 square foot

Check that the area is different.

Your rectangle has a length of ________ feet and a width of ________ feet.

Find the area.

________ feet × ________ feet = ________ square feet

A rectangle with the same perimeter but different area than Mark's poster has a length of ________ feet and a width of ________ feet.

Lesson Practice

Choose the correct answer.

Use the information for questions 1 and 2.

A rectangle has a length of 5 meters and a width of 8 meters.

1. What is the perimeter of the rectangle?

   ○ A. 26 m ○ C. 32 m
   ○ B. 30 m ○ D. 40 m

2. Which rectangle has the same perimeter as the rectangle above but a different area?

   ○ A.

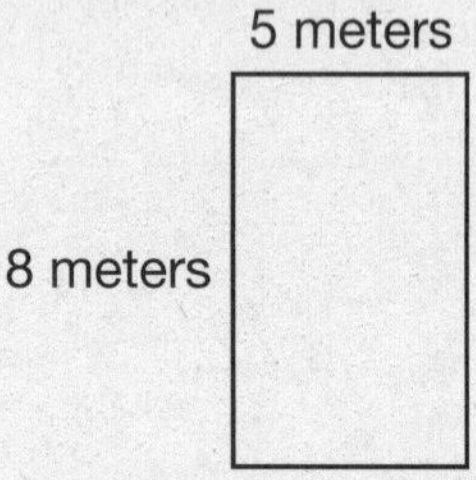

   ○ B.

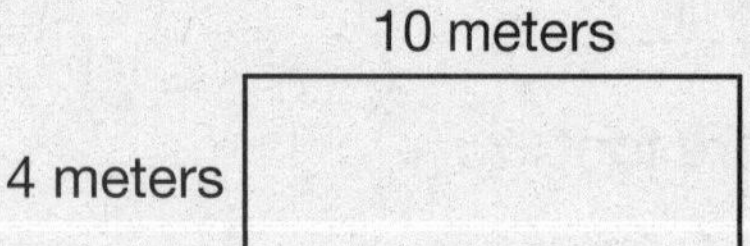

   ○ C.

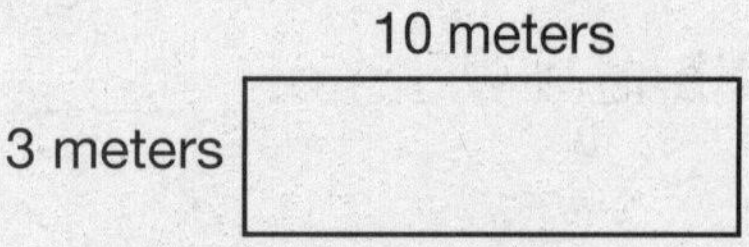

   ○ D.

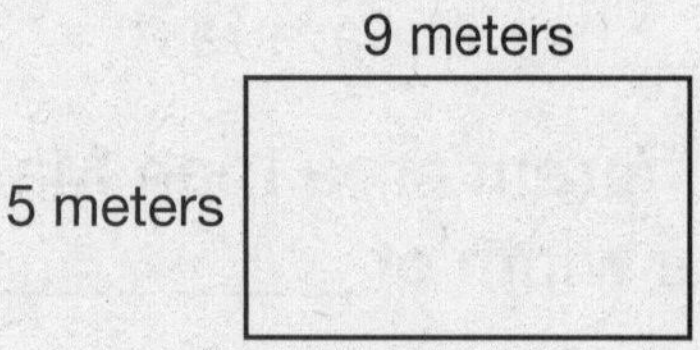

3. This rectangle has a length of 8 units and a width of 1 unit.

   Which of these rectangles has the same area, but a different perimeter?

   ○ A.

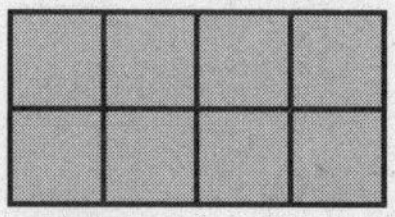

   ○ B.

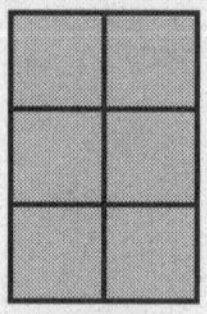

   ○ C.

   ○ D.

4. Which rectangle does **not** have the same perimeter as the others?

○ **A.** 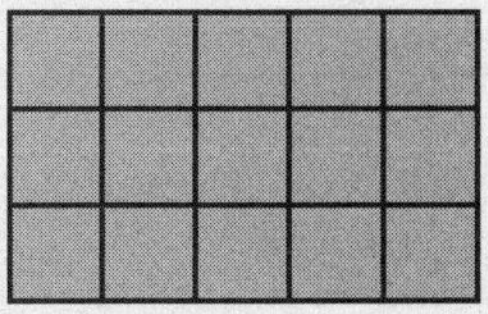

○ **C.** 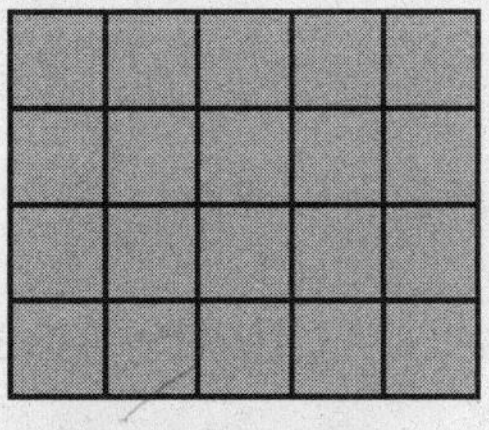

○ **B.** 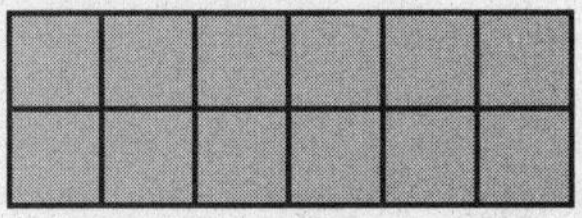

○ **D.** 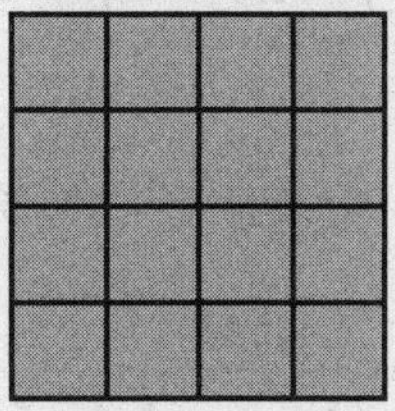

5. Ms. Jensen decides to build a garden with an area of 36 square meters.

A. In the grid below, draw two different rectangular gardens with the same area of 36 square meters.

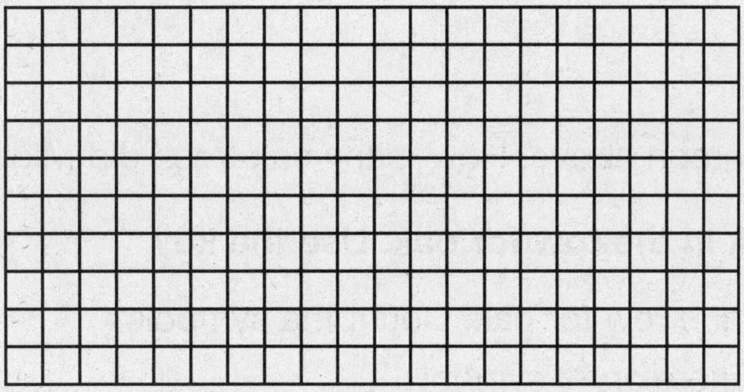

Key: □ = 1 square meter

B. What are the length and width of each garden that you drew in Part A? What is the perimeter of each garden?

Picture Graphs

Common Core State Standard:
3.MD.3

Getting the Idea

A **picture graph** uses pictures or symbols to display and compare **data.** The key tells how many each symbol represents.

Here are the parts of a picture graph.

This graph shows the types of trees Miguel saw last week.

title → **Type of Tree Miguel Saw**

Oak	♦ ♦ ♦ ♦ ♦ ♦ ♦ ♦ ♦
Pine	♦ ♦ ♦ ♦
Maple	♦ ♦ ♦ ♦ ♦ ← symbols
Elm	♦ ♦

Key: Each ♦ = 1 tree ← key

Example 1

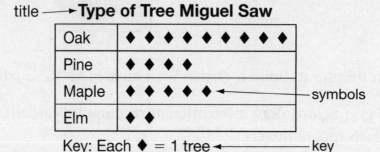

Use the picture graph above. How many oak trees did Miguel see?

Strategy **Look in the row for oak. Use the key.**

Step 1 Find the row for oak. Count the symbols.

There are 9 symbols.

Step 2 Use the key to find how many each symbol represents.

Each symbol equals 1 tree.

So 9 symbols equal 9 trees.

Solution **Miguel saw 9 oak trees.**

Example 2

The graph shows the number of books that some students read at the Read-A-Thon.

Books Read at the Read-A-Thon

Who read the greatest number of books? How many books did that student read?

Strategy **Find the row with the most symbols. Use the key.**

Step 1 Find the row with the most symbols.

The row for Emily shows 5 symbols.

Emily read the greatest number of books.

Step 2 Find the number of books Emily read.

Look at the key.

Each symbol equals 2 books.

Step 3 Use repeated addition or multiplication.

$2 + 2 + 2 + 2 + 2 = 10$ books

5 symbols $\times$ 2 books $=$ 10 books

Solution **Emily read the greatest number of books.**
She read 10 books.

Example 3

Monica asked some students about their favorite time of day. She recorded the data in the tally chart below.

Favorite Time of Day

Time of Day	Tally	Number of Votes
Morning	𝍸 𝍸	10
Afternoon	𝍸 I	6
Evening	𝍸 𝍸 IIII	14

Make a picture graph to show the same data.

Strategy **Choose a key that would work easily with the numbers in the table.**

Step 1 Make a chart with 3 rows for the picture graph.

Write the title above the picture graph.

Favorite Time of Day

Key: ________ = ____________

Step 2 Think of a picture to use for the key.

Use a ☺ to represent each student.

Step 3 Decide on a number to use for the key.

Each picture will represent 2 students.

Write the key at the bottom of the picture graph.

Step 4 Write the time of day in the first column.

Draw the correct number of symbols for each time of day.

Favorite Time of Day

Morning	☺ ☺ ☺ ☺ ☺
Afternoon	☺ ☺ ☺
Evening	☺ ☺ ☺ ☺ ☺ ☺ ☺

Key: Each ☺ = 2 students

Solution The picture graph is shown in Step 4.

Coached Example

Vera enjoys practicing the piano. The picture graph shows the time she spent practicing for 4 days in October.

Piano Practice

October 4	🕒 🕒
October 5	🕒 🕒 🕒 🕒
October 6	🕒 🕒 🕒 🕒 🕒 🕒
October 7	🕒 🕒 🕒 🕒 🕒 🕒 🕒 🕒

Key: Each 🕒 = 5 minutes

How many more minutes did Vera practice on October 6 than on October 4?

Look at the rows for October 4 and October 6.

There are ________ symbols for October 4.

There are ________ symbols for October 6.

There are ________ more symbols for October 6 than for October 4.

Look at the key. Each symbol represents ________ minutes.

Multiply the number of symbols by the number of minutes each symbol represents.

________ × ________ = ________

Vera practiced ________ more minutes on October 6 than on October 4.

Lesson Practice

Choose the correct answer.

Use the picture graph for questions 1–3.

Bags of Cookies Sold

Chocolate Chip	● ● ● ● ●
Oatmeal Raisin	● ●
Peanut Butter	● ● ●
Mint	● ● ● ●

Key: Each ● = 5 bags of cookies

1. What does each symbol in the picture graph represent?

- **A.** 5 jars
- **B.** 5 scoops
- **C.** 5 cups
- **D.** 5 bags

2. How many bags of mint cookies were sold?

- **A.** 20
- **B.** 16
- **C.** 9
- **D.** 4

3. How many fewer bags of oatmeal raisin cookies were sold than bags of chocolate chip cookies?

- **A.** 3
- **B.** 6
- **C.** 15
- **D.** 35

Use the picture graph for questions 4–6.

Students' Favorite Music

Pop	♫ ♫ ♫ ♫ ♫
Country	♫ ♫ ♫ ♫ ♫ ♫ ♫
Blues	♫ ♫ ♫ ♫
Classical	♫ ♫

Key: Each ♫ = 2 students

4. How many students chose country as their favorite type of music?

- **A.** 4
- **B.** 8
- **C.** 10
- **D.** 14

5. Which type of music did 8 students choose?

- **A.** pop
- **B.** country
- **C.** blues
- **D.** classical

6. R&B had 6 votes. What should the row for R&B look like?

A.

R & B	♫ ♫ ♫

B.

R & B	♫ ♫ ♫ ♫

C.

R & B	♫ ♫ ♫ ♫ ♫

D.

R & B	♫ ♫ ♫ ♫ ♫ ♫

Use the picture graph for questions 7 and 8.

Joyner's Hits in Fall Softball Season

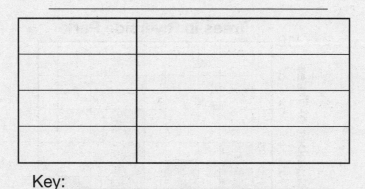

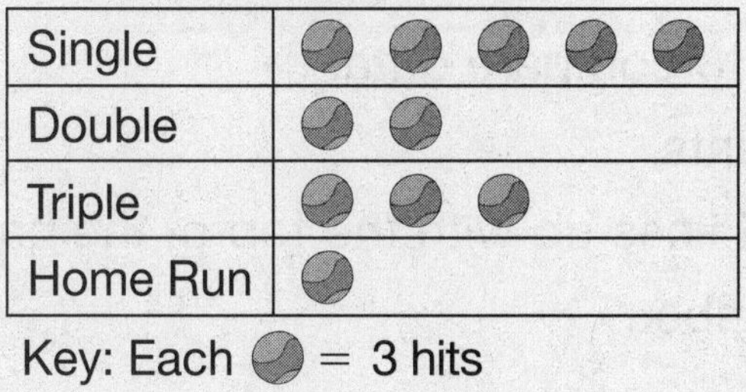

7. How many doubles did Joyner hit?

○ **A.** 2 ○ **C.** 6

○ **B.** 3 ○ **D.** 9

8. How many more singles than home runs did Joyner hit?

○ **A.** 4 ○ **C.** 12

○ **B.** 8 ○ **D.** 15

9. The table shows the animals that Michelle saw at a zoo.

Animals at a Zoo

Animal	Number of Animals
Bears	12
Elephants	6
Giraffes	10
Tigers	8

A. Make a picture graph of the information shown in the table.

____

Key: ______________________

B. How many more tigers than elephants did Michelle see at the zoo?

____

Common Core State Standard:
3.MD.3

Bar Graphs

Getting the Idea

A **bar graph** uses bars of different lengths to compare data.

The **scale** tells how many each bar represents.

To read the value of a bar, find the line that lines up with the top of the bar.

Follow that line to the scale to read the number.

Here are the parts of a bar graph.

This graph shows the number of students absent each day this week.

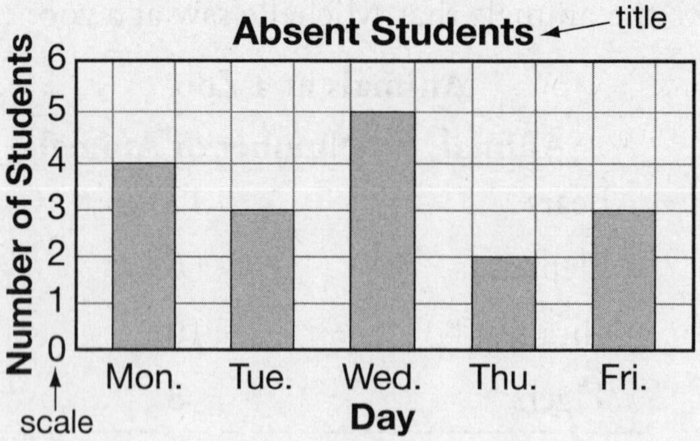

Example 1

This bar graph shows the number of different types of trees in Riverside Park.

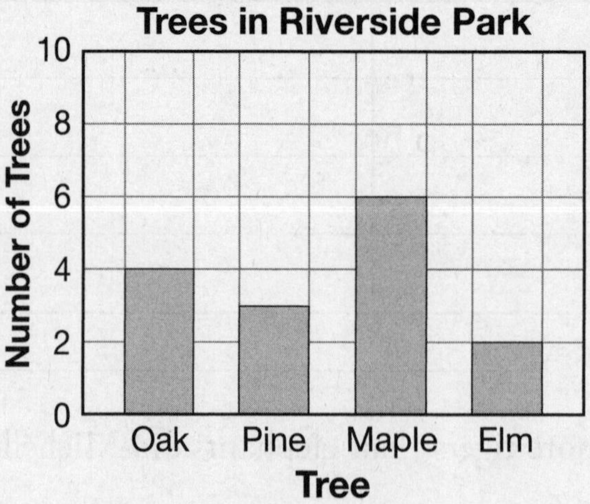

How many maple trees are in Riverside Park?

Strategy **Read the value for the bar for maple trees.**

Step 1 Find the bar for "maple" along the bottom of the graph.

Step 2 Look at the line that lines up with the top of the bar.

Move to the left and read the number on the scale.

It shows 6.

Solution **There are 6 maple trees in Riverside Park.**

Example 2

The bar graph shows the number of members in four different clubs.

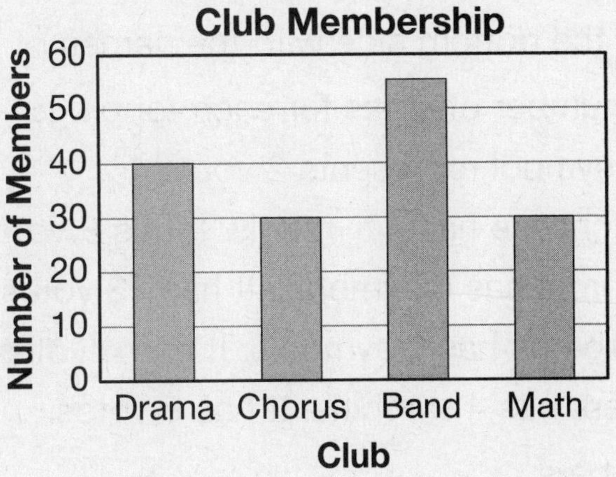

How many fewer members are in the math club than in the band?

Strategy **Find the number of members in each club.**

Step 1 Find the number of members in band.

The bar lines up halfway between 50 and 60.

The bar lines up with 55.

Step 2 Find the number of members in the math club.

The bar lines up with the number 30.

Step 3 Subtract to find the difference.

$55 - 30 = 25$

Solution **There are 25 fewer members in the math club than in band.**

Example 3

Isaac made this picture graph to show which pizza toppings his classmates prefer. Make a bar graph showing the same data.

Strategy **Find the value for each topping.**

Step 1 Label the graph.

Write the title: Favorite Pizza Toppings.

Label the side of the graph. Use a scale of 2.

Label the bottom side with the names of the toppings.

Step 2 Find the number of votes for each topping.

Each symbol represents 2 votes.

Extra cheese has 4 symbols. It has 8 votes.

Pepperoni has 6 symbols. It has 12 votes.

Mushrooms has 2 symbols. It has 4 votes.

Peppers has 4 symbols. It has 8 votes.

Step 3 Draw the bars.

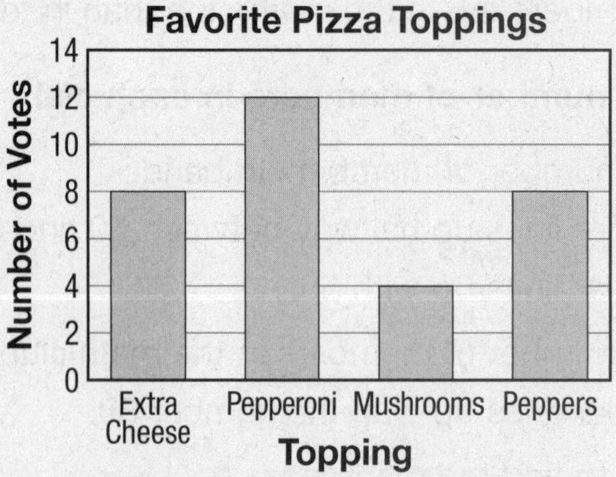

Solution **The bar graph is shown in Step 3.**

Coached Example

Students in the third grade voted on a school mascot.

School Mascot Vote

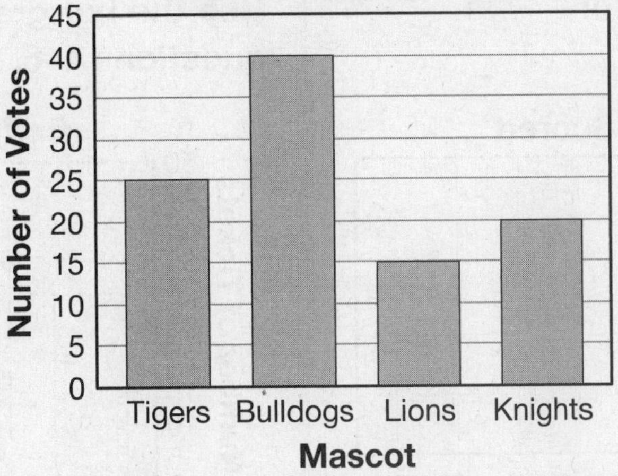

How many more students voted for Bulldogs than for Lions?

Find the number of votes for each mascot.

Find the number of votes for Bulldogs.

The bar lines up with ________.

Find the number of votes for Lions.

The bar lines up with ________.

Subtract.

________ – ________ = ________

So, ________ more students voted for Bulldogs than for Lions.

Lesson Practice

Choose the correct answer.

Use the bar graph for questions 1–3.

1. Who scored 4 goals?

   ○ **A.** Chelsea ○ **C.** Sandy
   ○ **B.** Maurice ○ **D.** Adam

2. Who scored more than 6 goals but fewer than 10 goals?

   ○ **A.** Chelsea ○ **C.** Sandy
   ○ **B.** Maurice ○ **D.** Adam

3. How many goals did Maurice and Adam score in all?

   ○ **A.** 16 ○ **C.** 10
   ○ **B.** 12 ○ **D.** 5

Use the bar graph for questions 4–6.

4. Who sold exactly 35 tickets?

   ○ **A.** Mike ○ **C.** Lynn
   ○ **B.** Joel ○ **D.** Rita

5. How many more tickets did Rita sell than Mike?

   ○ **A.** 15 ○ **C.** 45
   ○ **B.** 30 ○ **D.** 75

6. How many tickets did Joel and Lynn sell in all?

   ○ **A.** 85 ○ **C.** 40
   ○ **B.** 75 ○ **D.** 35

Use the bar graph for questions 7 and 8.

Eric asked some students about which states they would most like to visit. The graph shows his data.

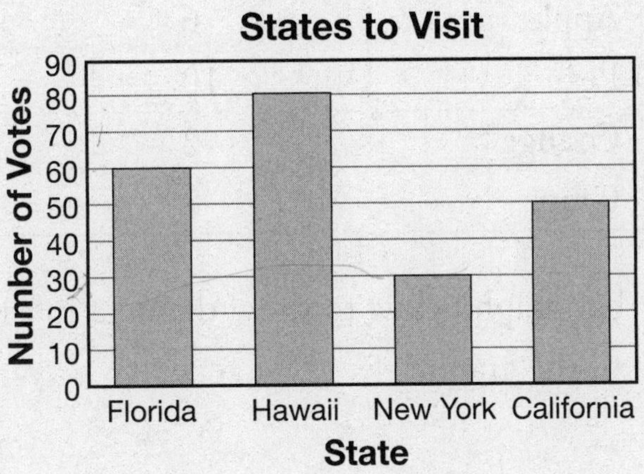

7. How many students voted for Florida?

- ○ **A.** 30
- ○ **B.** 50
- ● **C.** 60
- ○ **D.** 80

8. How many fewer students voted for New York than Hawaii?

- ○ **A.** 30
- ○ **B.** 40
- ● **C.** 50
- ○ **D.** 80

9. The table shows students' votes for their favorite fruits.

Favorite Fruits

Fruit	Number of Students
Apple	6
Pear	10
Orange	2
Grape	8

A. Complete the bar graph below to show the data in the table. Label all parts of the graph.

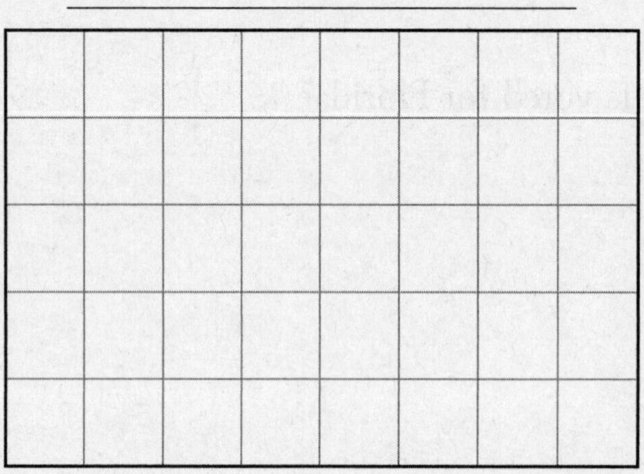

B. How many more students chose pears as their favorite fruit than apples?

____

Common Core State Standard: 3.MD.4

Measure Lengths

Getting the Idea

Length is the measure of how long, wide, or tall something is.

An **inch (in.)** is a unit of length in the customary system.

You can use a ruler to measure the length of small objects.

This inch ruler has 8 equal spaces between the numbered marks for each inch.

There are 4 equal spaces between the marks for every $\frac{1}{2}$ inch.

There are 2 equal spaces between the marks for every $\frac{1}{4}$ inch.

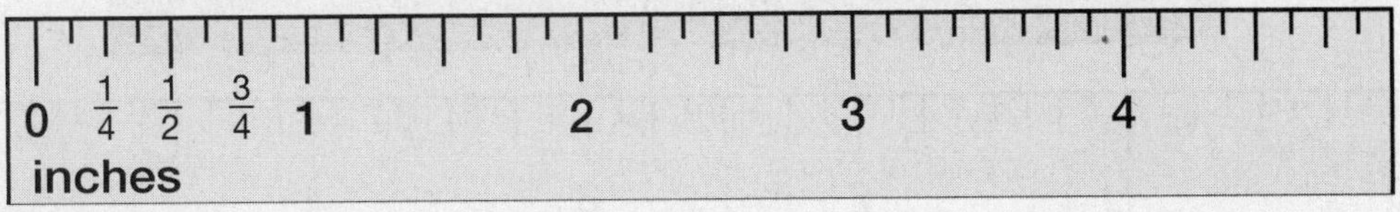

Example 1

To the nearest $\frac{1}{2}$ inch, what is the length of this crayon?

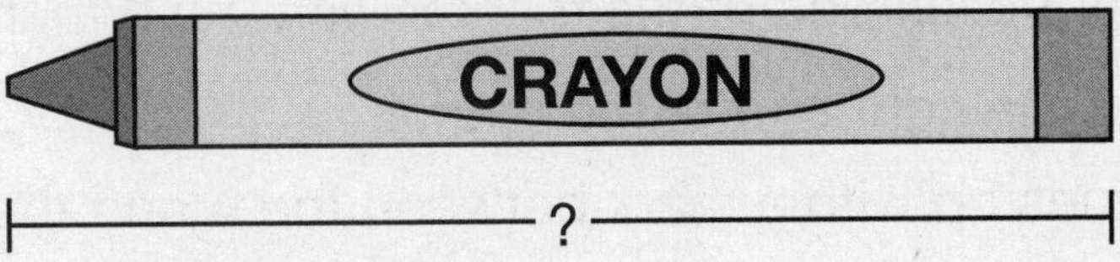

Strategy **Use an inch ruler.**

Step 1 Line up the left end of the crayon with the 0 mark on the ruler.

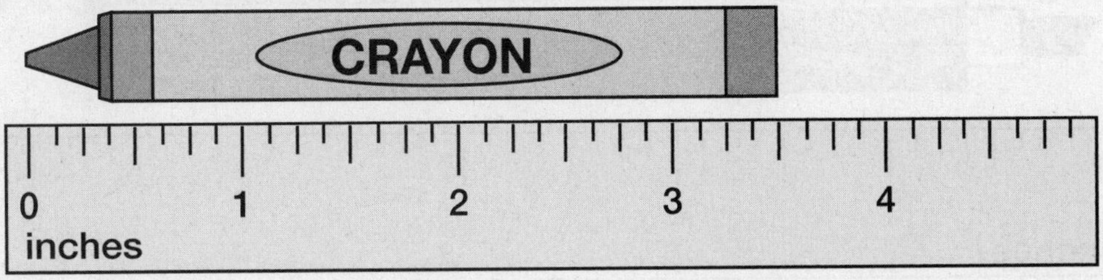

Step 2 Look at the right end of the crayon.

It lines up with a mark between the 3 and the 4.

The crayon is more than 3 inches long.

Step 3 Count the spaces past the 3-inch mark.

The mark is 4 spaces after the 3-inch mark.

So the crayon is $3\frac{1}{2}$ inches long.

Solution **The length of the crayon is $3\frac{1}{2}$ inches long.**

Example 2

To the nearest $\frac{1}{4}$ inch, what is the length of the marker?

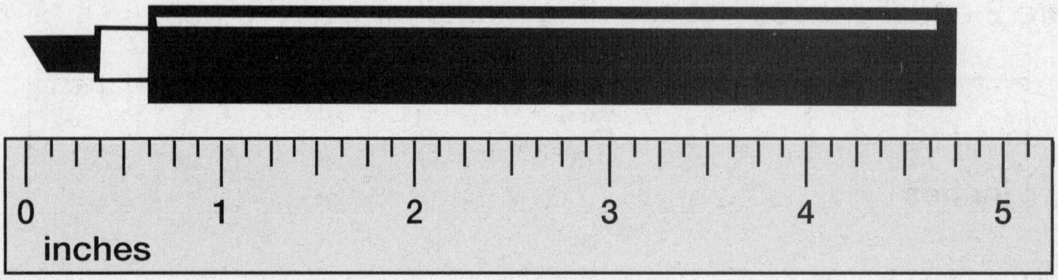

Strategy **Use an inch ruler.**

Step 1 Line up the left end of the marker with the 0 mark on the ruler.

Step 2 Look at the right end of the marker.

It lines up with a mark between the 4 and the 5.

The marker is more than 4 inches long.

Step 3 Count the spaces past the 4-inch mark.

The marker is $4\frac{3}{4}$ inches long.

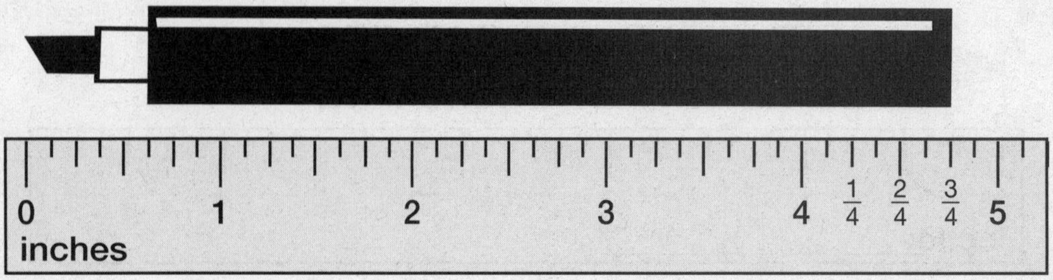

Solution **The length of the marker is $4\frac{3}{4}$ inches.**

Millimeters (mm) and **centimeters (cm)** are two units of length in the metric system.

10 millimeters are equal to 1 centimeter.

You can use a centimeter ruler to measure lengths in millimeters.

This ruler has 10 equal spaces between the numbered marks for each centimeter.

There is 1 equal space between the millimeter marks.

Since 10 millimeters are equal to 1 centimeter, 1 millimeter is $\frac{1}{10}$ centimeter.

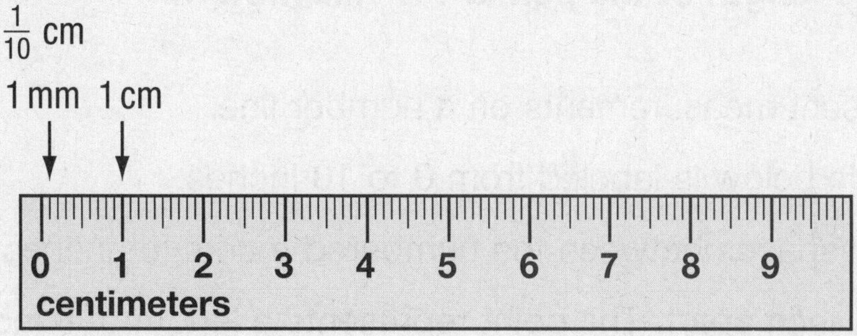

Example 3

What is the length of this pen in millimeters?

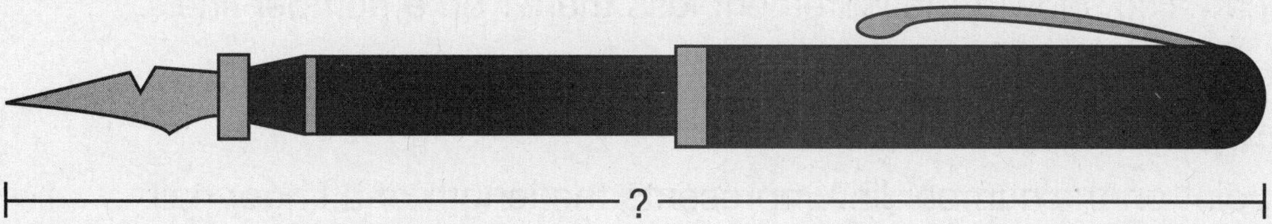

Strategy **Use a centimeter ruler.**

Step 1 Line up the left end of the pen with the 0 mark on the ruler.

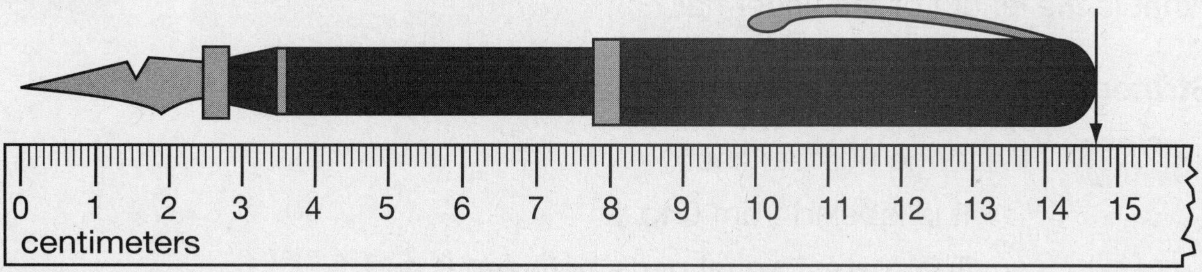

Step 2 Look at the right end of the pen.

It lines up with a mark between 14 and 15.

The pen is more than 14 centimeters long.

14 centimeters = 140 millimeters

So the pen is more than 140 millimeters long.

Step 3 Count the spaces past the 140-millimeter mark.

The mark is 7 spaces after the 140-millimeter mark.

The pen is 147 millimeters long.

Solution **The length of the pen is 147 millimeters.**

You can represent measurements on a number line.

The number line below is labeled from 0 to 10 inches.

It has 10 equal spaces between the numbered marks for inches.

Each mark is 1 inch apart. The point represents a length of 8 inches.

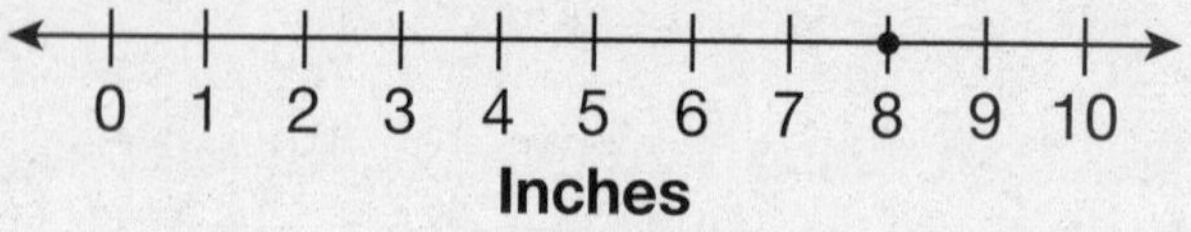

You can also show a measurement less than 1 on a number line.

Example 4

The point on the number line represents the length of a finger nail.

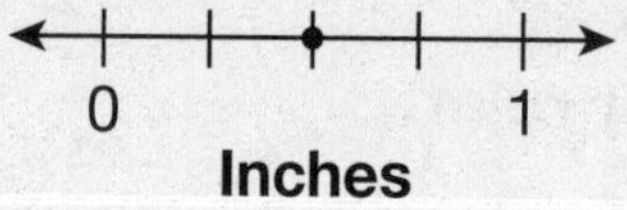

What is the length of the finger nail?

Strategy **Count the number of equal parts between 0 and 1.**

Step 1 Read the number line.

It is labeled from 0 to 1.

There are 4 equal parts between 0 and 1.

Step 2 Decide what each mark represents.

Since there are 4 equal parts, each mark is $\frac{1}{4}$.

Step 3 Find the value of the point.

The point is on the 2nd mark after the 0 mark.

So the point is on $\frac{2}{4}$ inch.

Solution **The length of the finger nail is $\frac{2}{4}$ inch.**

In Example 4, notice that $\frac{2}{4}$ is in the middle between 0 and 1.

So $\frac{2}{4}$ inch is the same as $\frac{1}{2}$ inch.

Coached Example

The point on the number line represents the length of a small worm.

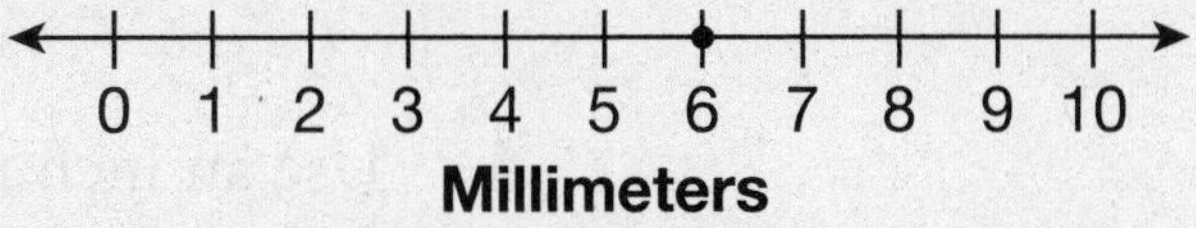

What is the length of the worm?

Read the number line.

The number line is labeled from ________ to ________.

What are the units of the numbers? ________________________

What does each mark represent? ________________________

Find the value of the point.

The point is on the ________ mark after the 0 mark.

So the point is on ________ millimeters.

The length of the worm is __________ millimeters.

Lesson Practice

Choose the correct answer.

1. Use an inch ruler. To the nearest $\frac{1}{4}$ inch, what is the length of the scissors?

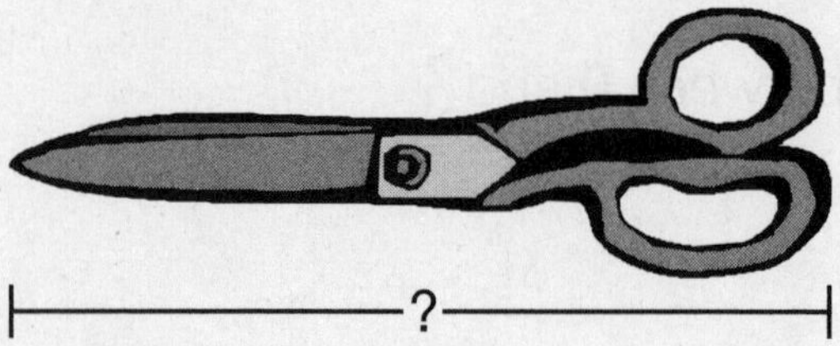

○ **A.** $2\frac{1}{4}$ inches

○ **B.** $2\frac{1}{2}$ inches

○ **C.** $2\frac{3}{4}$ inches

○ **D.** $3\frac{1}{4}$ inches

2. Use a centimeter ruler. What is the length of this ribbon?

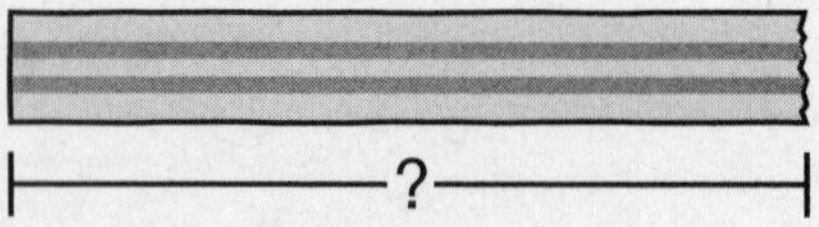

○ **A.** 41 millimeters

○ **B.** 55 millimeters

○ **C.** 59 millimeters

○ **D.** 62 millimeters

3. Use a centimeter ruler. What is the length of this eraser?

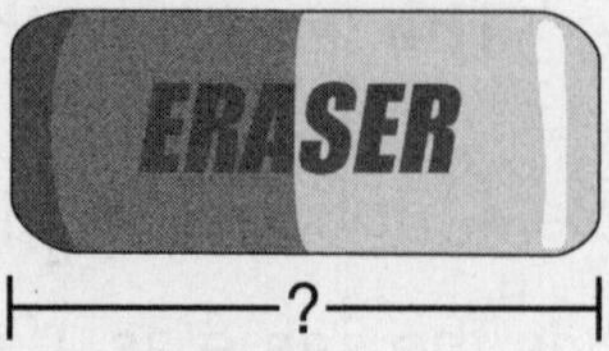

○ **A.** 2 centimeters

○ **B.** 4 centimeters

○ **C.** 6 centimeters

○ **D.** 7 centimeters

4. Use an inch ruler. To the nearest $\frac{1}{4}$ inch, what is the length of the key?

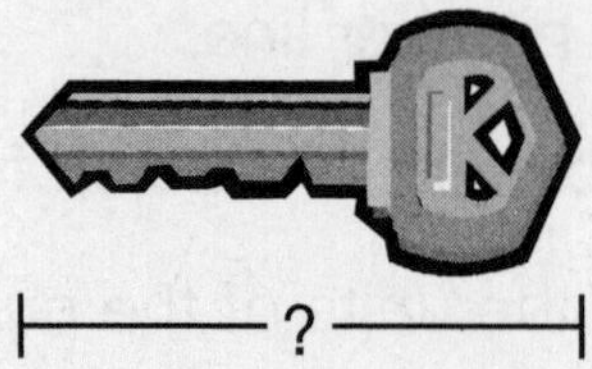

○ **A.** $1\frac{1}{4}$ inches

○ **B.** $1\frac{1}{2}$ inches

○ **C.** $2\frac{1}{4}$ inches

○ **D.** $2\frac{1}{2}$ inches

5. The point on the number line represents the length of a card.

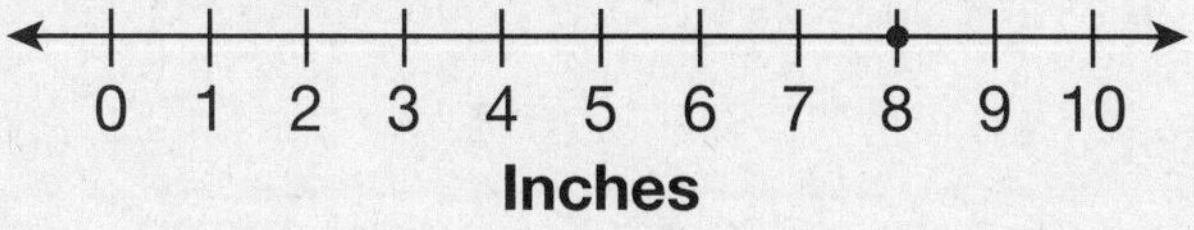

What is the length of the card?

○ **A.** $\frac{1}{4}$ inch

○ **B.** $\frac{1}{2}$ inch

○ **C.** 4 inches

○ **D.** 8 inches

6. The point on the number line represents the length of a small sticker.

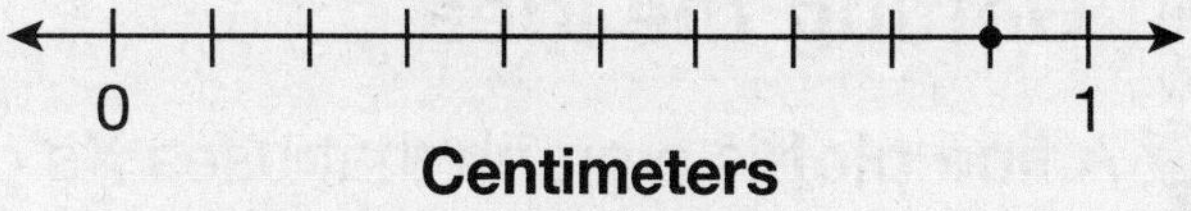

What is the length of the sticker?

○ **A.** $\frac{6}{10}$ centimeter

○ **B.** $\frac{9}{10}$ centimeter

○ **C.** 1 centimeter

○ **D.** 9 centimeters

7. Victor drew an arrow on a poster.

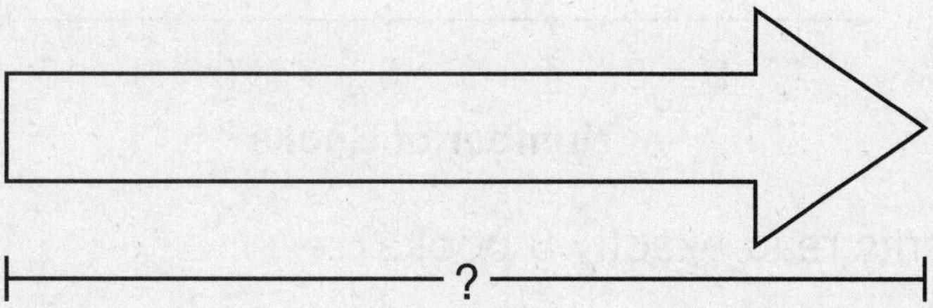

A. Use an inch ruler. To the nearest $\frac{1}{2}$ inch, what is the length of the arrow?

____

B. Use a centimeter ruler. To the nearest centimeter, what is the length of the arrow?

____

Common Core State Standard:
3.MD.4

Line Plots

Getting the Idea

A **line plot** is a graph that uses Xs or dots above a number line to record data. To read a line plot, count the number of Xs or dots above the number on the number line.

Example 1

The line plot shows the number of books each student in Stephen's class read.

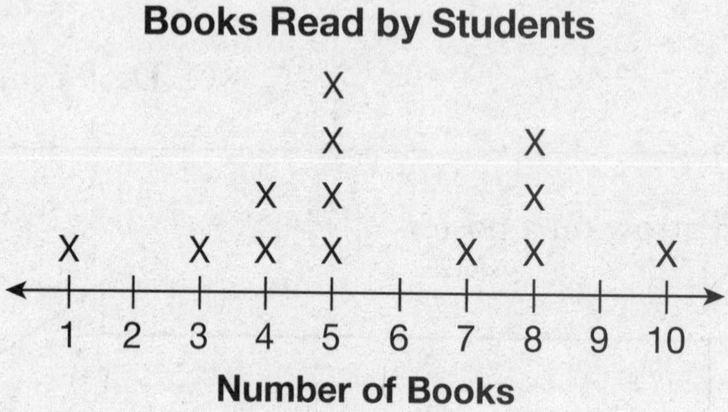

How many students read exactly 8 books?

Strategy **Count the number of Xs above the number 8.**

Step 1 Understand the line plot.

The number line shows the number of books.

Each X represents 1 student.

Step 2 Count the number of Xs above the number 8.

There are 3 Xs above the number 8.

So, 3 students read 8 books.

Solution **Three students read exactly 8 books.**

Some line plots show fractions, such as in measurements.

Example 2

Kevin measured some strings and recorded the data in the line plot below.

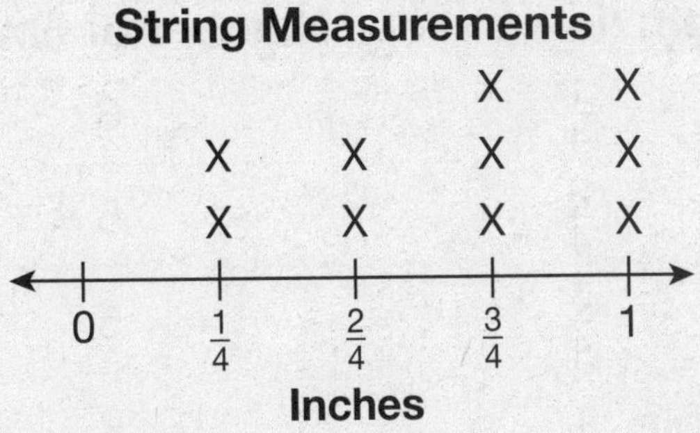

How many strings are $\frac{1}{4}$-inch long?

Strategy **Count the number of Xs above the fraction $\frac{1}{4}$.**

Step 1 Understand the line plot.

The number line shows lengths from 0 to 1 inch.

The Xs represent the number of strings.

Step 2 Count the number of Xs above the $\frac{1}{4}$-inch mark.

There are 2 Xs above the $\frac{1}{4}$-inch mark.

So, 2 strings are $\frac{1}{4}$-inch long.

Solution **Two strings are $\frac{1}{4}$-inch long.**

Example 3

Tia measured the widths of some people's index fingers. She made a table of the data. Make a line plot to show the data in the table.

Index Finger Widths

Width (in inches)	Number of Fingers
$\frac{1}{4}$	3
$\frac{2}{4}$	5
$\frac{3}{4}$	5
1	2

Strategy **Make a number line. Record the data in the plot.**

Step 1 Look at the widths.

The widths are $\frac{1}{4}$ inch, $\frac{2}{4}$ inch, $\frac{3}{4}$ inch, and 1 inch.

Step 2 Make a number line from 0 to 1 in fourths.

Label the number line.

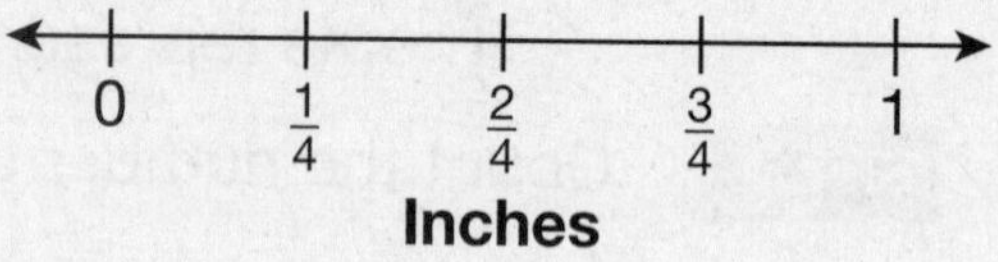

Step 3 Find the number of fingers for each measurement.

$\frac{1}{4}$-inch width – 3

$\frac{2}{4}$-inch width – 5

$\frac{3}{4}$-inch width – 5

1-inch width – 2

Step 4 Draw an X to represent each finger above each measurement.

Write a title for the line plot.

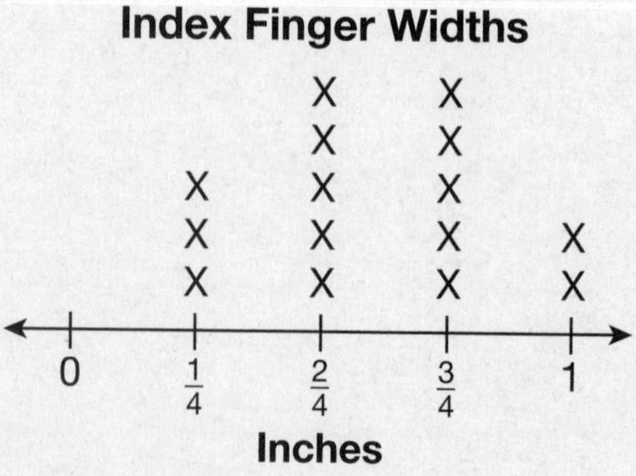

Solution **The line plot is shown in Step 4.**

Coached Example

Shanice measured some buttons and recorded the data in the line plot below.

Button Measurements

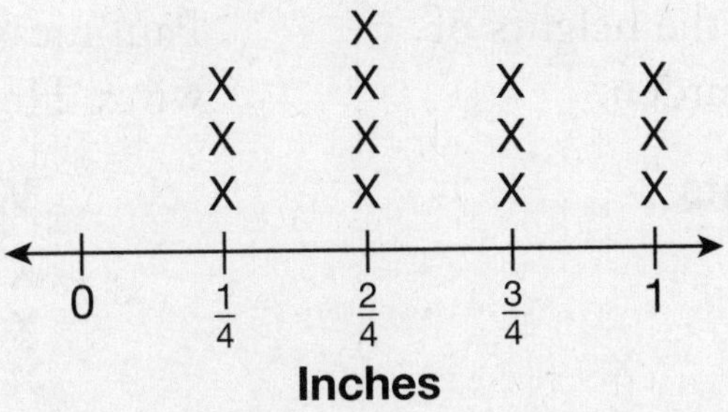

How many buttons are $\frac{2}{4}$ inch long?

Understand the line plot.

The number line shows the lengths in ______________.

The Xs represent the number of ______________.

Find the number of buttons that are $\frac{2}{4}$ inch long.

Count the number of Xs above the fraction __________ on the number line.

There are __________ Xs above $\frac{2}{4}$ inch.

So, __________ buttons are $\frac{2}{4}$ inch long.

Lesson Practice

Choose the correct answer.

Use the line plot for questions 1–3.

The line plot shows the heights of plants in Marissa's garden.

Plant Heights

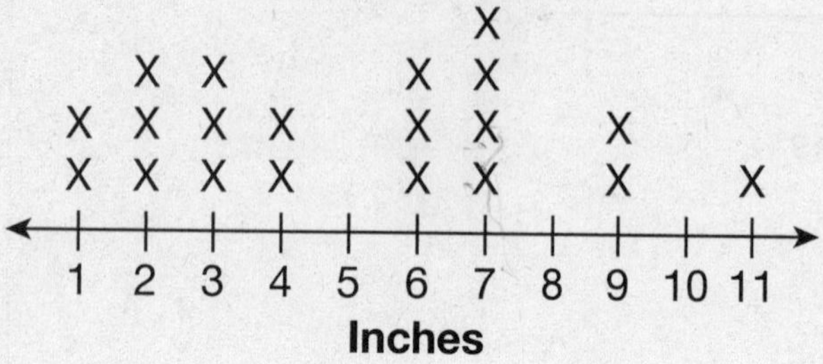

1. How many plants are 7 inches tall?

   ○ **A.** 0
   ○ **B.** 2
   ○ **C.** 3
   ○ **D.** 4

2. How tall is the tallest plant in Marissa's garden?

   ○ **A.** 10 inches
   ○ **B.** 11 inches
   ○ **C.** 12 inches
   ○ **D.** 15 inches

3. How many more plants are 6 inches tall than 9 inches tall?

   ○ **A.** 1
   ○ **B.** 2
   ○ **C.** 3
   ○ **D.** 4

Use the line plot for questions 4–6.

Paul measured the lengths of some wires. He made a line plot.

Wire Lengths

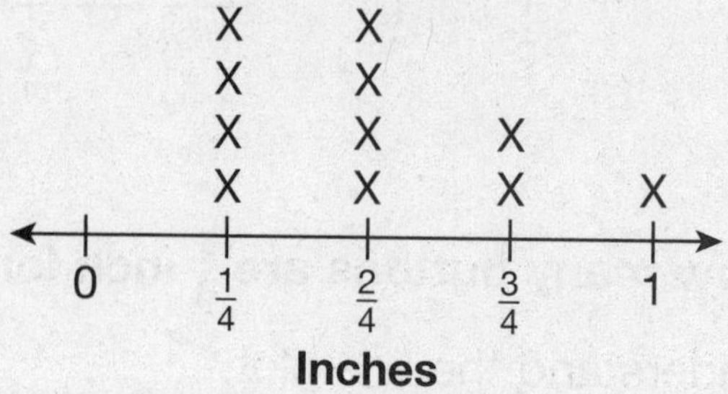

4. How many wires are $\frac{3}{4}$ inch or longer?

   ○ **A.** 1
   ○ **B.** 2
   ○ **C.** 3
   ○ **D.** 4

5. How many more wires are $\frac{2}{4}$ inch than 1 inch?

   ○ **A.** 1
   ○ **B.** 3
   ○ **C.** 4
   ○ **D.** 5

6. How many wires did Paul measure in all?

   ○ **A.** 4
   ○ **B.** 8
   ○ **C.** 10
   ○ **D.** 11

7. Jess measured some flower petals. She made a table of the data.

Flower Petal Lengths

Length (in inches)	Number of Petals
$\frac{1}{2}$	5
1	8

A. Make a line plot to show the data in the table. Be sure to label the title.

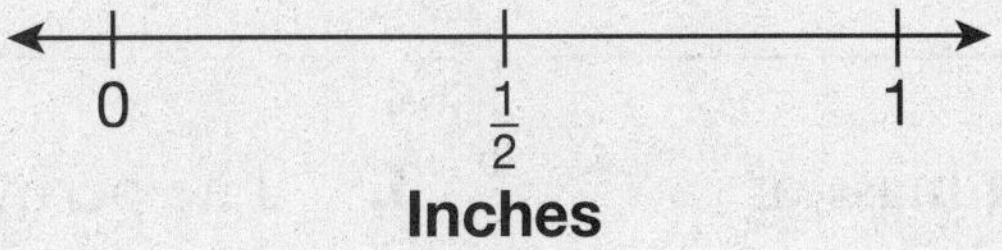

B. How many more 1-inch petals are there than $\frac{1}{2}$-inch petals? Show your work.

Domain 4: Cumulative Assessment for Lessons 23–33

1. Yesterday, Gary did his homework before dinner. He worked on math problems for 30 minutes and studied spelling for 28 minutes.

How many minutes did Gary spend on yesterday's homework?

- ○ **A.** 2 minutes
- ○ **B.** 12 minutes
- ○ **C.** 18 minutes
- ○ **D.** 58 minutes

2. A russet potato has a mass of 375 grams. A sweet potato has a mass of 245 grams. How much greater is the mass of the russet potato than the sweet potato?

- ○ **A.** 620 grams
- ○ **B.** 420 grams
- ○ **C.** 130 grams
- ○ **D.** 30 grams

3. The perimeter of this triangle is 42 inches. What is the missing side length?

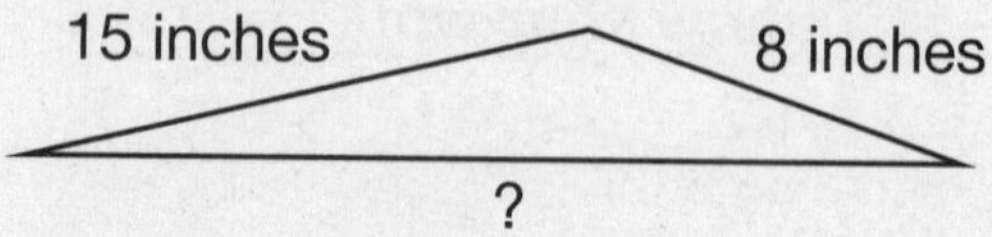

- ○ **A.** 17 inches
- ○ **B.** 19 inches
- ○ **C.** 23 inches
- ○ **D.** 29 inches

4. What is the area of the shaded figure?

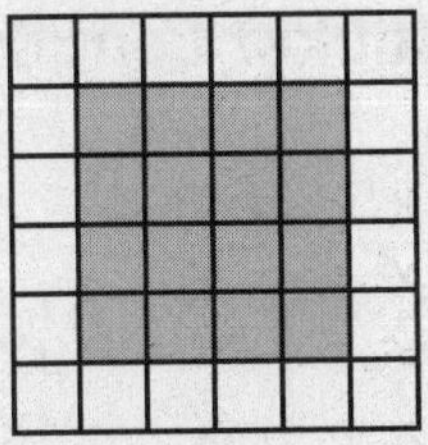

Key: ▢ = 1 square foot

○ **A.** 4 square feet
○ **B.** 16 square feet
○ **C.** 18 square feet
○ **D.** 20 square feet

5. Jade cut a rectangle with an area of 8 square inches. Which figure did Jade cut?

○ **A.**

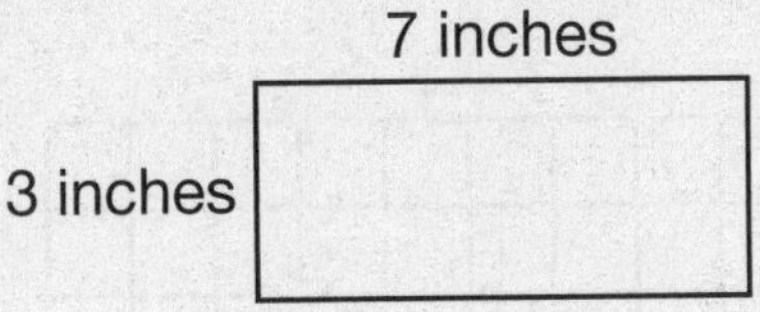

○ **B.**

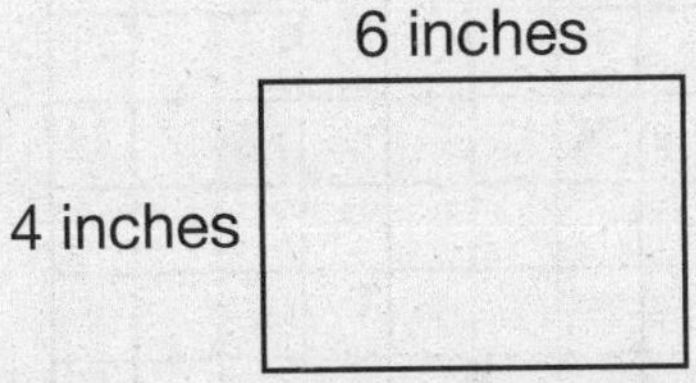

○ **C.**

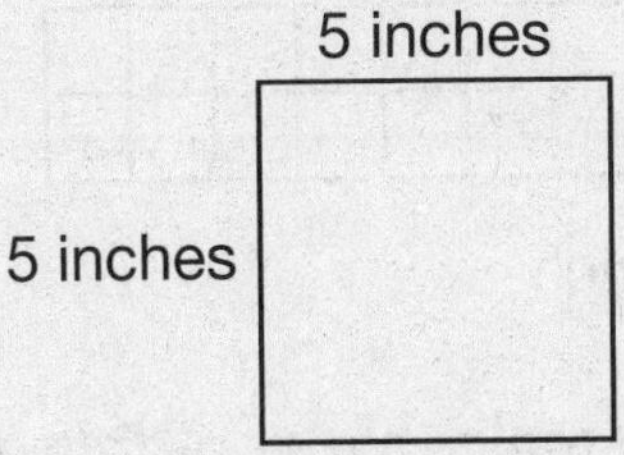

○ **D.**

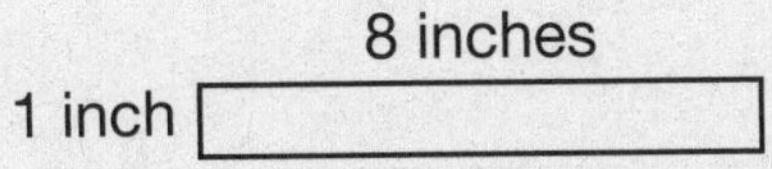

6. The bar graph shows the number of animals at a wildlife preserve.

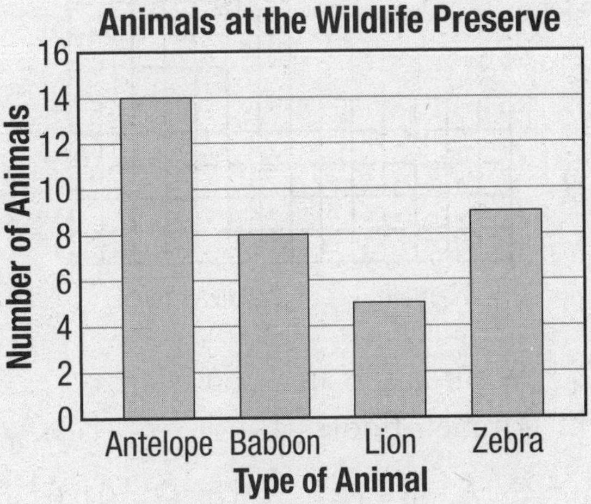

How many fewer lions than baboons are at the wildlife preserve?

○ **A.** 1
○ **B.** 3
○ **C.** 6
○ **D.** 13

7. A basketball player drank water before the game, at halftime, and at the end of the game. She drank 200 milliliters of water each time. How much water did she drink in all?

○ **A.** 200 milliliters
○ **B.** 400 milliliters
○ **C.** 600 milliliters
○ **D.** 800 milliliters

8. Vivian wants to install new tiles for her kitchen floor.

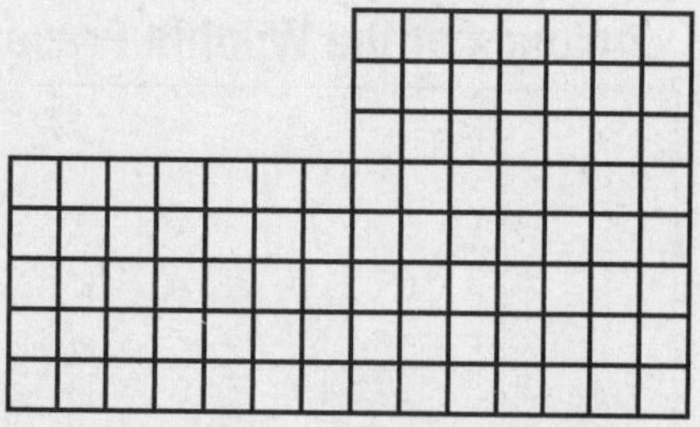

Key: ☐ = 1 square foot

What is the area of Vivian's kitchen floor?

○ **A.** 83 ft^2 ○ **C.** 91 ft^2

○ **B.** 84 ft^2 ○ **D.** 98 ft^2

9. The line plot shows the lengths of some stickers.

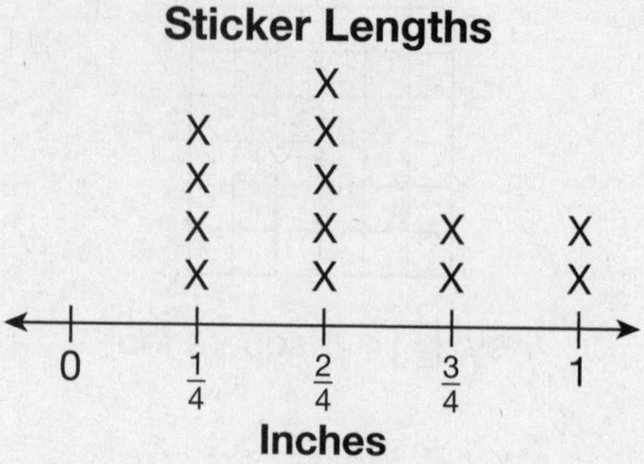

How many more $\frac{2}{4}$-inch stickers are there than 1-inch stickers?

____

10. Angelina wants to cut a rectangle that has a perimeter of 18 inches.

A. On the grid below, draw two different rectangles with a perimeter of 18 inches.

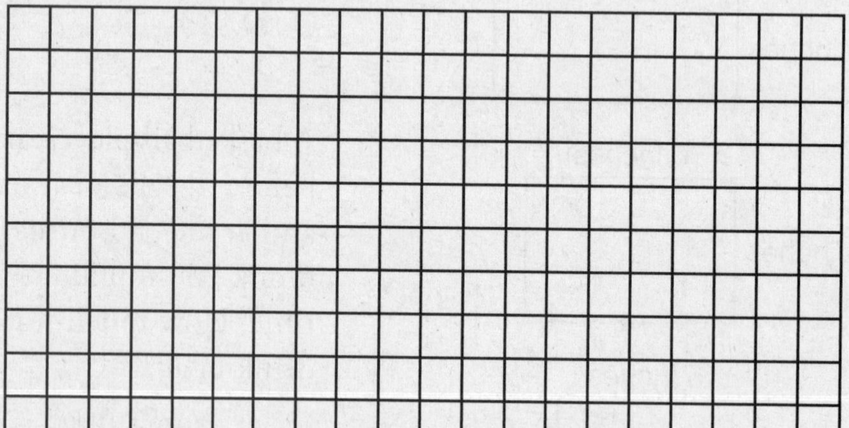

Key: ☐ = 1 square inch

B. What are the length and width of each rectangle that you drew for Part A? What is the area of each rectangle?

____

____

____

Domain 5

Geometry

Domain 5: Diagnostic Assessment for Lessons 34–36

Lesson 34 Two-Dimensional Shapes
3.G.1

Lesson 35 Quadrilaterals
3.G.1

Lesson 36 Area of Shapes
3.G.2

Domain 5: Cumulative Assessment for Lessons 34–36

Domain 5: Diagnostic Assessment for Lessons 34–36

1. Which sentence is true about the pentagon?

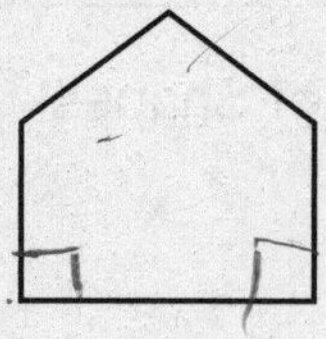

- ○ **A.** It has 6 angles.
- ○ **B.** It has 5 straight sides.
- ○ **C.** It has 5 square corners.
- ○ **D.** It has 6 straight sides.

2. Which two-dimensional shape has no angles?

- ○ **A.** triangle
- ○ **B.** rectangle
- ○ **C.** rhombus
- ○ **D.** circle

3. Which of the following is a quadrilateral?

- ○ **A.** circle
- ○ **B.** triangle
- ○ **C.** trapezoid
- ○ **D.** hexagon

4. Which describes the area of the shaded part of the rectangle?

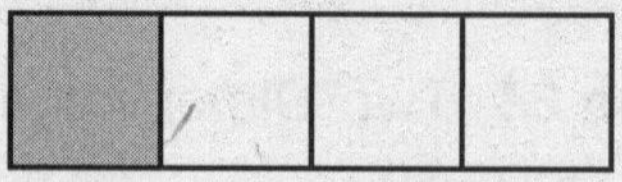

Key: □ = 1 square centimeter

- ○ **A.** $\frac{1}{3}$ of the total area
- ○ **B.** $\frac{1}{4}$ of the total area
- ○ **C.** $\frac{1}{6}$ of the total area
- ○ **D.** $\frac{1}{8}$ of the total area

5. Which is **not** a quadrilateral?

- ○ **A.** octagon
- ○ **B.** square
- ○ **C.** rhombus
- ○ **D.** parallelogram

6. Which is a trapezoid?

○ **A.** 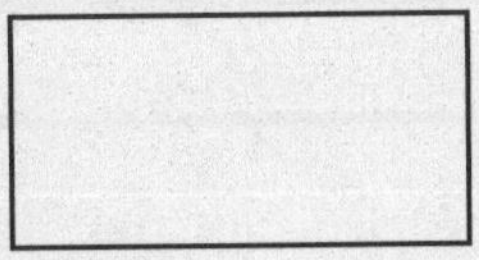

○ **B.** 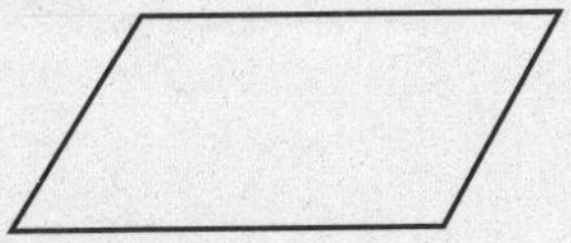

○ **C.** 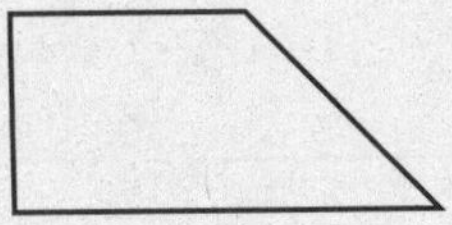

○ **D.** 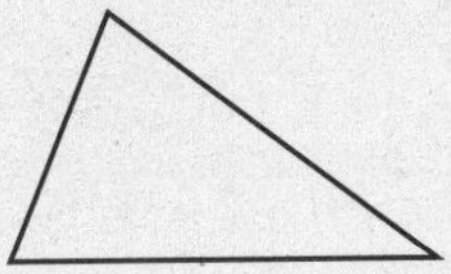

Use the rectangle below for questions 7 and 8.

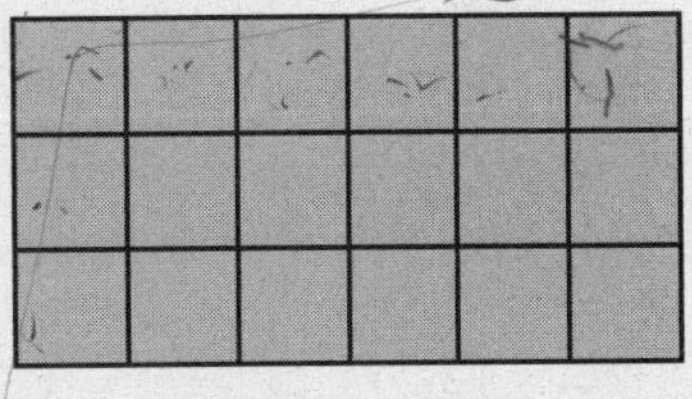

Key: ▒ = 1 square inch

7. Which can be used to find the area of the rectangle?

○ **A.** 3 + 3 + 3 = ☐

○ **B.** 6 + 6 + 6 = ☐

○ **C.** 6 + 3 + 6 = ☐

○ **D.** 3 + 6 + 9 = ☐

8. What is the area of the rectangle?

○ **A.** 9 square inches

○ **B.** 12 square inches

○ **C.** 18 square inches

○ **D.** 29 square inches

9. The square below has 4 equal parts.

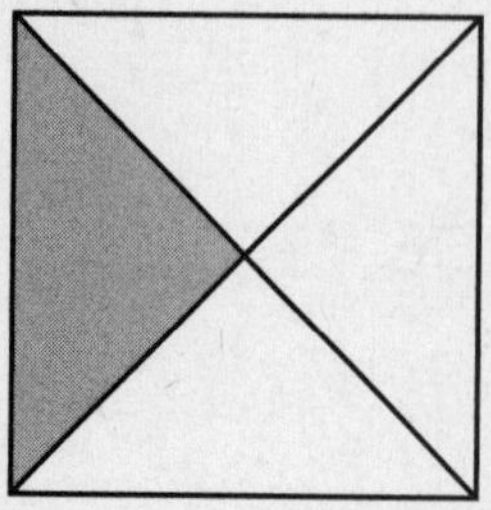

What fraction describes the area of the shaded part?

____

10. Kim made this shape using triangles.

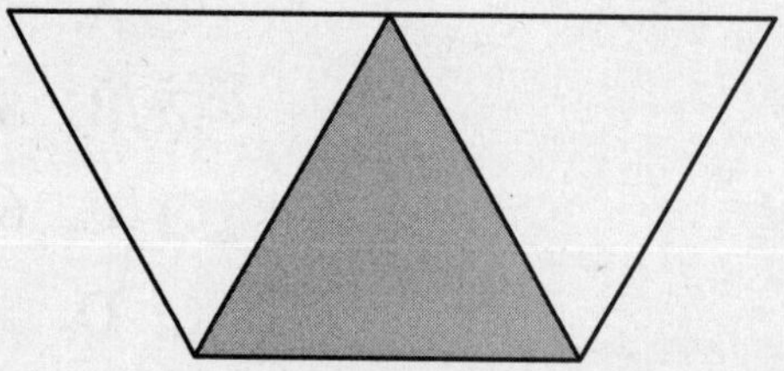

A. Which shape did Kim make? Explain your answer.

____

____

B. What fraction describes the area of the shaded part of the shape? Explain your answer.

____

____

Domain 5 • Lesson 34

Common Core State Standard:
3.G.1

Two-Dimensional Shapes

Getting the Idea

A **two-dimensional shape** is a flat shape that has length and width.

A **circle** is a two-dimensional shape with a curved side. It has no corners or edges.

Polygons are two-dimensional shapes with straight sides. The polygons below are named by the numbers of **sides** and **angles** they have.

A **triangle** is a polygon with 3 sides and 3 angles.

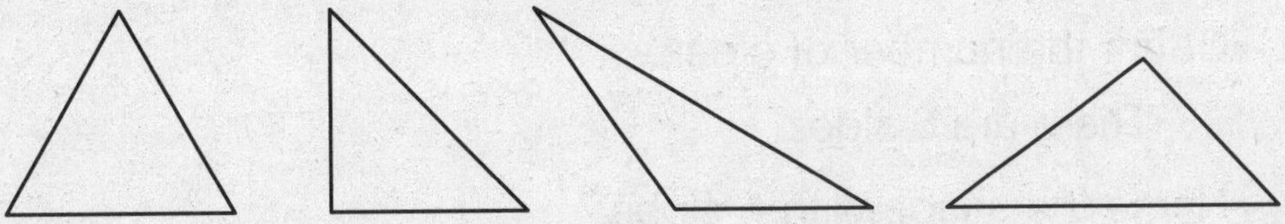

A **quadrilateral** is a polygon with 4 sides and 4 angles.

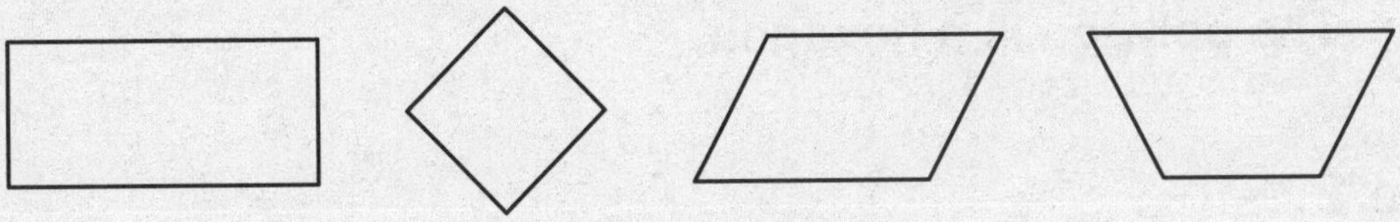

A **pentagon** is a polygon with 5 sides and 5 angles.

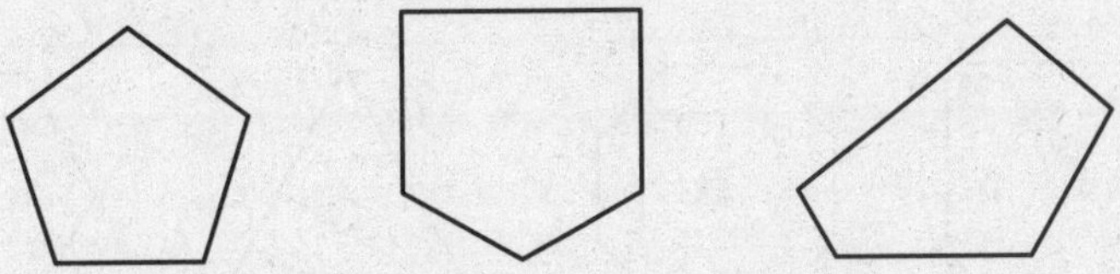

A **hexagon** is a polygon with 6 sides and 6 angles.

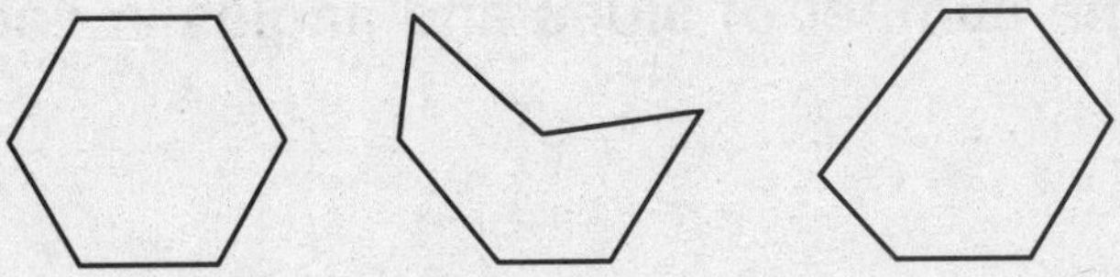

An **octagon** is a polygon with 8 sides and 8 angles.

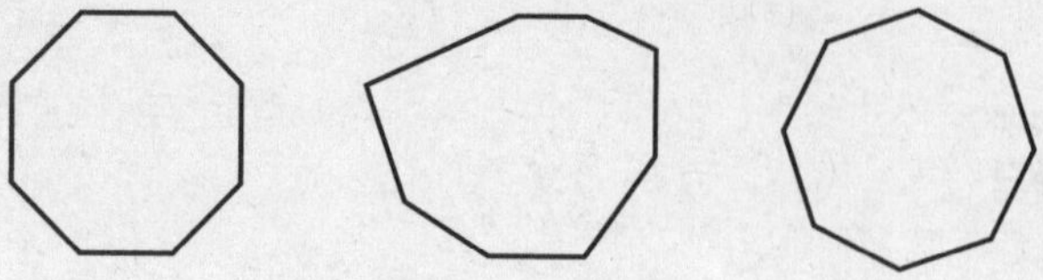

Example 1

What is the name of this polygon?

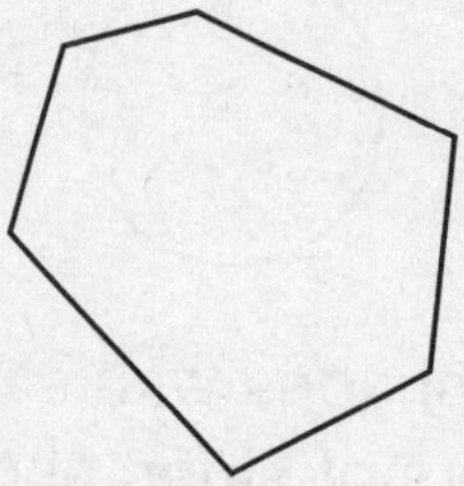

Strategy **Find the number of sides.**

Step 1 Count the number of sides.

There are 6 sides.

Step 2 Name the shape with 6 sides.

A hexagon has 6 sides.

Solution **The polygon is a hexagon.**

Example 2

Carlos drew a shape. It had more angles than a pentagon and fewer sides than an octagon. Which shape did Carlos draw?

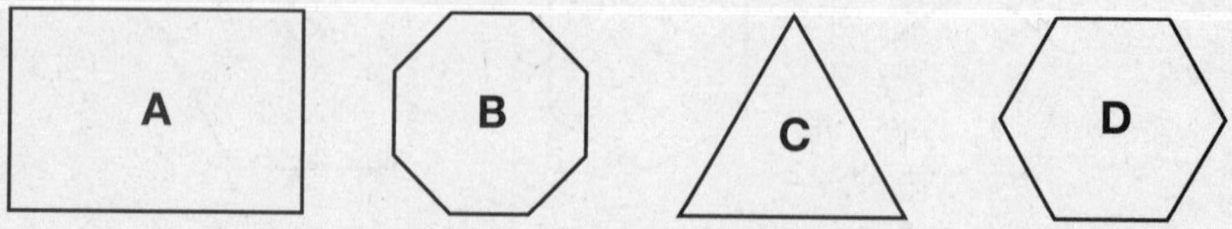

Strategy **Count the number of sides and angles in each shape.**

Step 1 Think about the clues.

The shape has more angles than a pentagon.

A pentagon has 5 angles, so the shape has more than 5 angles.

The shape has fewer sides than an octagon.

An octagon has 8 sides, so the shape has less than 8 sides.

Step 2 Review the shape in each choice.

Shape A has 4 sides and 4 angles. It is a rectangle.

Shape B has 8 sides and 8 angles. It is an octagon.

Shape C has 3 sides and 3 angles. It is a triangle.

Shape D has 6 sides and 6 angles. It is a hexagon.

Step 3 Choose the correct shape.

Look for the shape that has more than 5 angles and less than 8 sides.

A hexagon has more than 5 angles and less than 8 sides.

Solution **Carlos drew a hexagon.**

Example 3

Andrea drew some quadrilaterals on the cover of her art book.

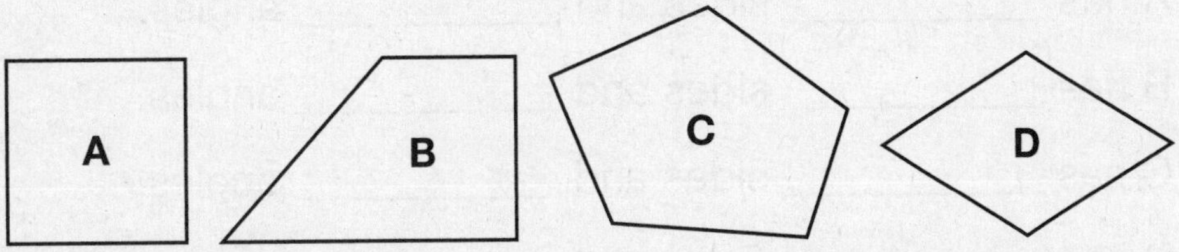

Which shape could **not** be one of the shapes Andrea drew?

Strategy **Count the number of sides and angles in each shape.**

Step 1 Review the shape in each choice.

Shape A has 4 sides and 4 angles.

Shape B has 4 sides and 4 angles.

Shape C has 5 sides and 5 angles.

Shape D has 4 sides and 4 angles.

Step 2 Think about a quadrilateral.

A quadrilateral is a polygon with 4 sides and 4 angles.

Step 3 Pick the shape that is not a quadrilateral.

Shape C is not a quadrilateral.

Solution **Shape C could not be a shape that Andrea drew.**

Coached Example

Which shape is an octagon?

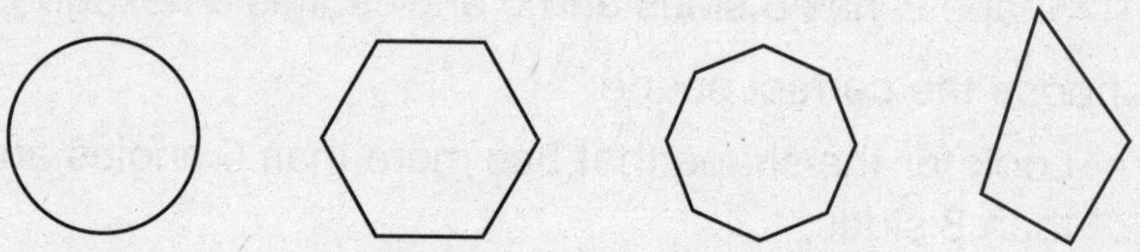

Shape A Shape B Shape C Shape D

Think about an octagon.

An octagon has __________ sides and __________ angles.

Review the shape in each choice.

Shape A has __________ sides and __________ angles.

Shape B has __________ sides and __________ angles.

Shape C has __________ sides and __________ angles.

Shape D has __________ sides and __________ angles.

Shape __________ is an octagon.

Lesson Practice

Choose the correct answer.

1. What is the name of the shape below?

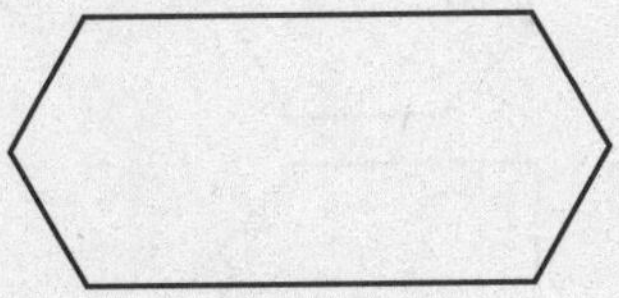

- ○ **A.** triangle
- ○ **B.** quadrilateral
- ○ **C.** hexagon
- ○ **D.** octagon

2. Which two-dimensional shape is **not** a polygon?

- ○ **A.** triangle
- ○ **B.** quadrilateral
- ○ **C.** circle
- ○ **D.** octagon

3. Which shape has 1 less side than a quadrilateral?

- ○ **A.** triangle
- ○ **B.** circle
- ○ **C.** pentagon
- ○ **D.** hexagon

4. Which is **not** a triangle?

○ **A.**

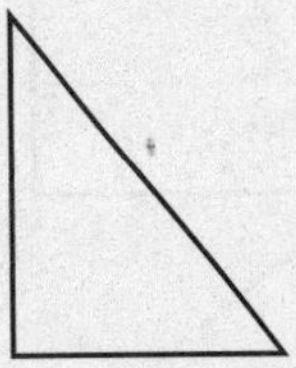

○ **B.**

○ **C.**

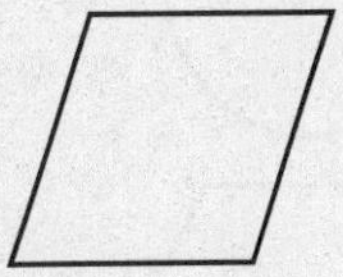

○ **D.**

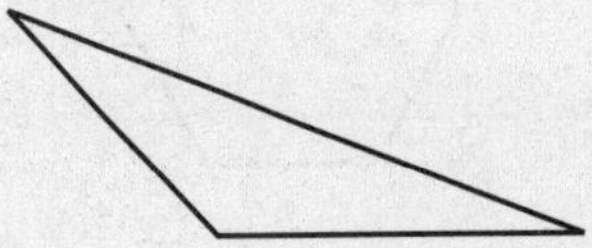

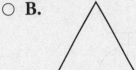

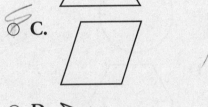

5. What is the name of this polygon?

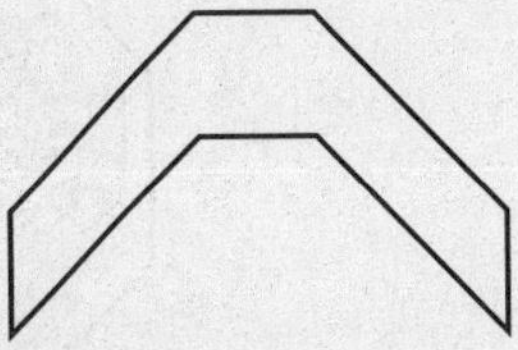

- ○ **A.** quadrilateral
- ○ **B.** pentagon
- ○ **C.** hexagon
- ○ **D.** octagon

6. Allison drew a shape that has more angles than a triangle, but fewer angles than a pentagon. Which shape did Allison draw?

○ **A.** 

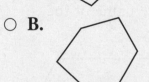

○ **B.** 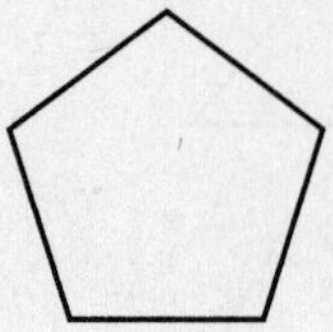

○ **C.** 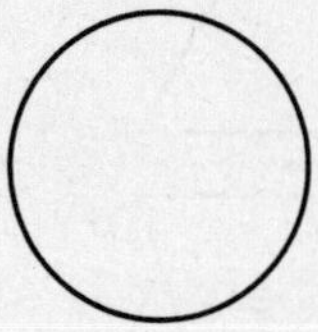

○ **D.** 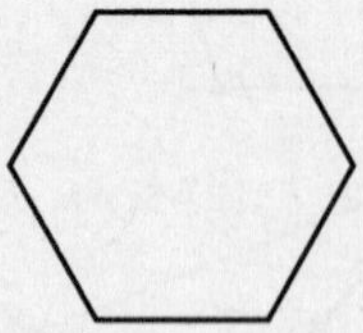

7. Which shape is a hexagon?

○ **A.** 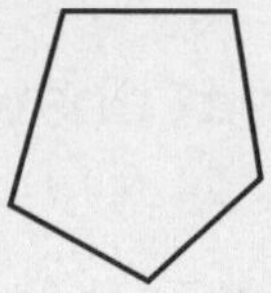

○ **B.** 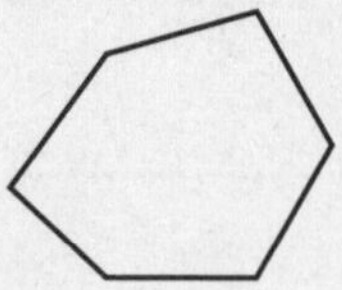

○ **C.** 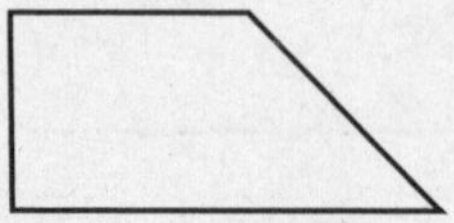

○ **D.** 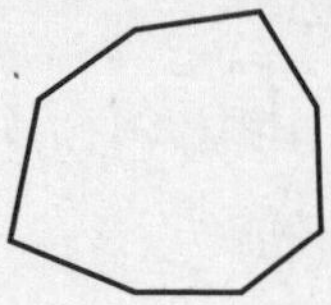

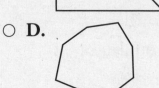

8. Look at these two street signs.

A. How many sides and angles does each street sign have? What is the name of the shape of each sign?

____

____

B. Which sign has more sides? How many more sides?

____

Quadrilaterals

Common Core State Standard:
3.G.1

Getting the Idea

A **quadrilateral** is a polygon with 4 sides and 4 angles.
You can sort quadrilaterals by their side lengths and angles.

These quadrilaterals have at least one pair of opposite sides that are **parallel**. Parallel sides remain the same distance apart and never meet.

Quadrilaterals

Name	Example	Characteristics
Parallelogram	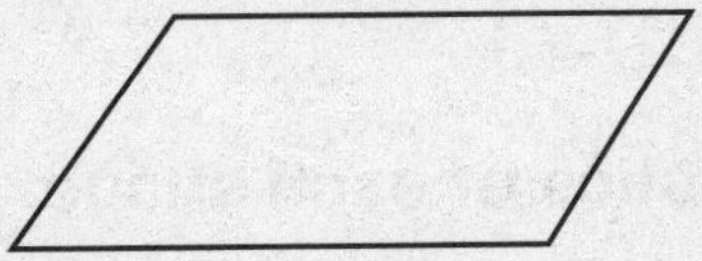	Opposite sides have the same length. Both pairs of opposite sides are parallel.
Rectangle	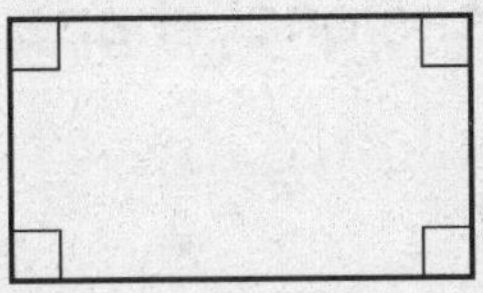	Opposite sides have the same length. Both pairs of opposite sides are parallel. The 4 angles are square corners.
Square	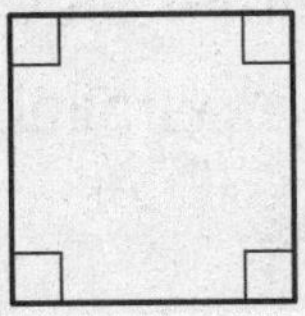	All sides have the same length. Both pairs of opposite sides are parallel. The 4 angles are square corners.
Rhombus	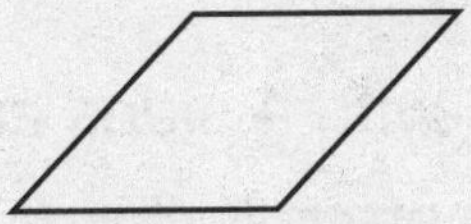	All sides have the same length. Both pairs of opposite sides are parallel.
Trapezoid	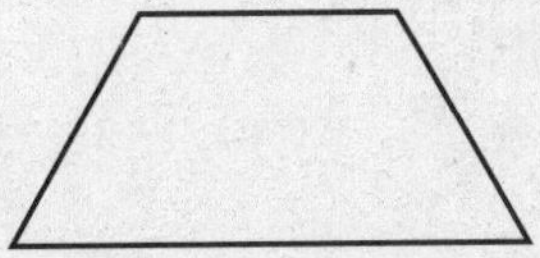	Exactly 1 pair of sides is parallel.

Example 1

Which is the best name for this shape?

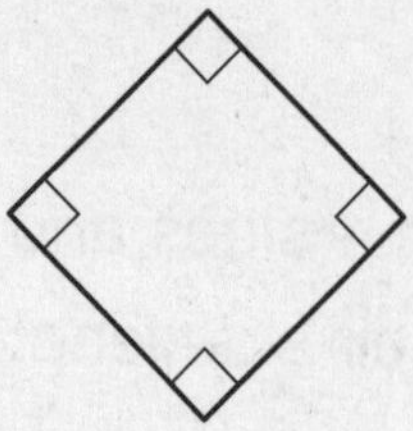

○ **A.** quadrilateral

○ **B.** rectangle

○ **C.** rhombus

○ **D.** square

Strategy **Think about the characteristics of each shape.**

Step 1 Think about a quadrilateral.

A quadrilateral is a two-dimensional shape with 4 sides and 4 angles.

This shape is a quadrilateral.

Step 2 Think about a rectangle.

A rectangle is a quadrilateral with 4 sides and 4 angles.

Both pairs of opposite sides are parallel.

The 4 angles are square corners.

This shape is a rectangle.

Step 3 Think about a rhombus.

A rhombus is a quadrilateral with 4 sides and 4 angles.

Both pairs of opposite sides are parallel.

All sides have the same length.

This shape is a rhombus.

Step 4 Think about a square.

A square is a quadrilateral with 4 sides and 4 angles.

Both pairs of opposite sides are parallel.

All sides have the same length.

The 4 angles are square corners.

This shape is a square.

This is the best name for the shape.

Solution **The best name for this shape is square, Choice D.**

Example 2

What is the name for this shape?

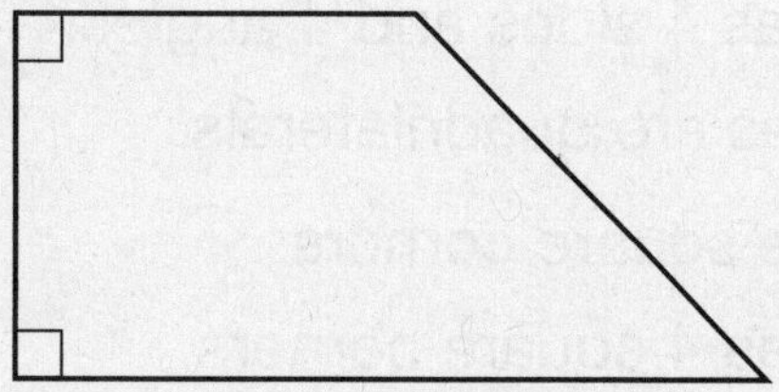

Strategy **Look at the angles and the sides of the shape.**

Step 1 Decide if the shape is a quadrilateral.

The shape has 4 straight sides and 4 angles.

It is a quadrilateral.

Step 2 Look at the angles of the shape.

The shape has 2 square corners.

So, the shape is not a square or a rectangle.

Step 3 Look at the sides of the shape.

The shape has only 1 pair of parallel sides.

So, the shape is not a parallelogram. It is a trapezoid.

Solution **The name for this shape is trapezoid.**

Example 3

How are these shapes alike? How are these shapes different? Name each shape.

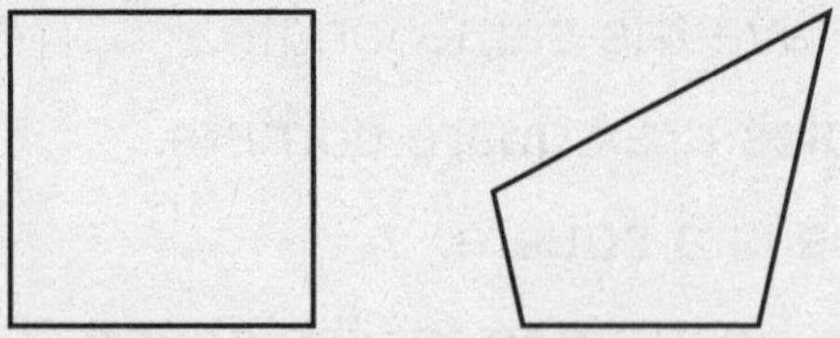

Strategy **Compare the sides and angles.**

Step 1 Count the number of sides and angles.

Shape A has 4 sides and 4 angles.

Shape B has 4 sides and 4 angles.

Both shapes are quadrilaterals.

Step 2 See if there are square corners.

Shape A has 4 square corners.

Shape B does not have any square corners.

Step 3 Look at the sides.

Shape A has 4 equal sides.

Both pairs of opposite sides are parallel.

Shape B has no equal sides and no parallel sides.

Step 4 Name the shapes.

Shape A is a square.

Shape B is a quadrilateral.

Solution **The shapes are alike because both are quadrilaterals. The shapes are different because Shape A has 2 pairs of parallel sides and 4 square corners. Shape A is a square. Shape B is a quadrilateral.**

Coached Example

Which shape is a rectangle, but is not a square?

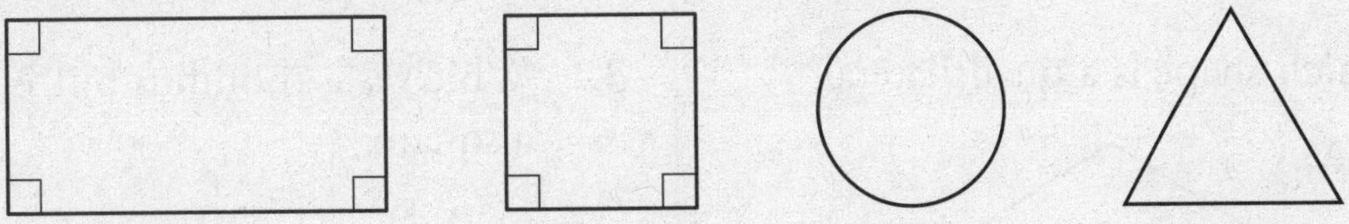

Shape F Shape G Shape H Shape J

Think about a rectangle.

A rectangle has ___________ sides and ___________ square corners.

The opposite sides of a rectangle have the ____________________ length.

Both pairs of opposite sides of a rectangle are ____________________.

Think about a square.

A square is a special rectangle with 4 ____________________ sides.

Shape ___________ and Shape ___________ have 4 square corners.

Only Shape ___________ does not have equal sides.

Shape ___________ is a rectangle but is not a square.

Lesson Practice

Choose the correct answer.

1. Which shape is a quadrilateral?

○ **A.**

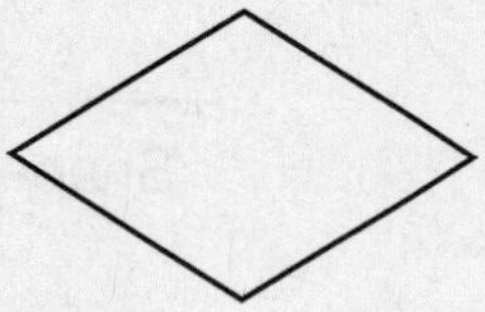

○ **B.**

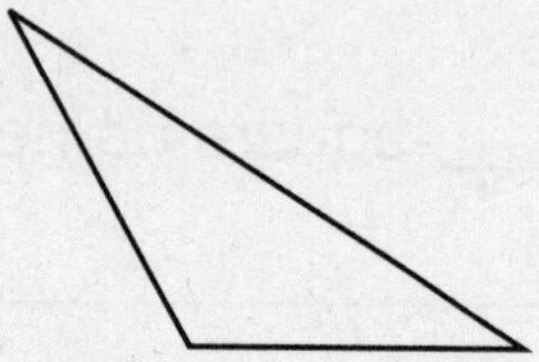

○ **C.**

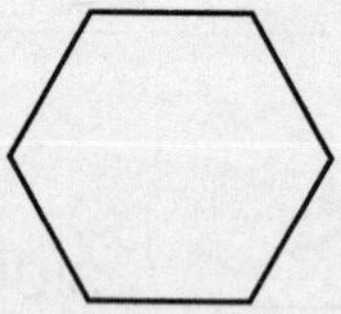

○ **D.**

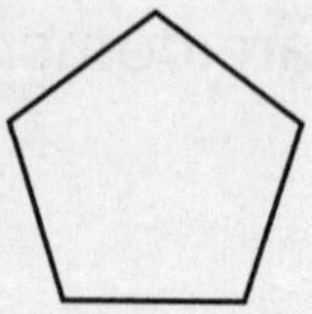

2. Which quadrilateral is shown below?

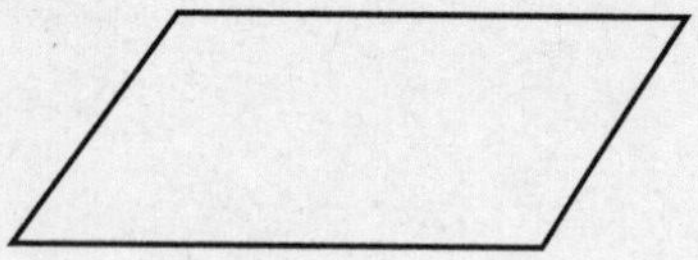

○ **A.** rectangle

○ **B.** parallelogram

○ **C.** trapezoid

○ **D.** rhombus

3. Which is a rhombus but is **not** a square?

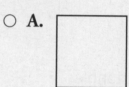

○ **A.**

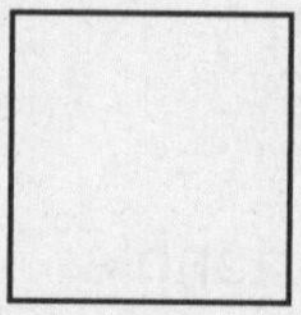

○ **B.**

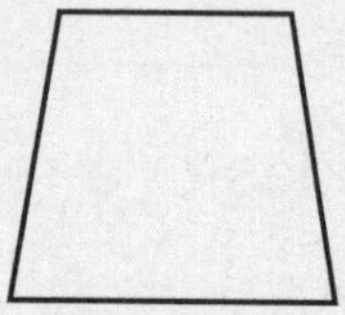

○ **C.**

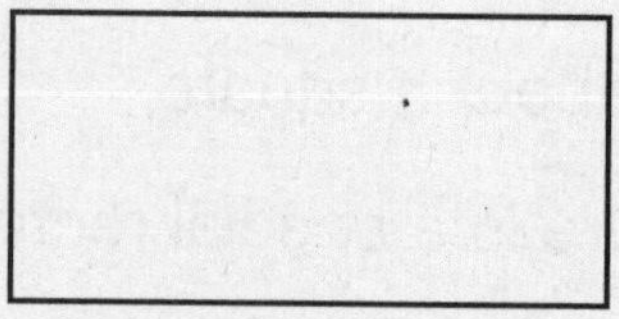

○ **D.**

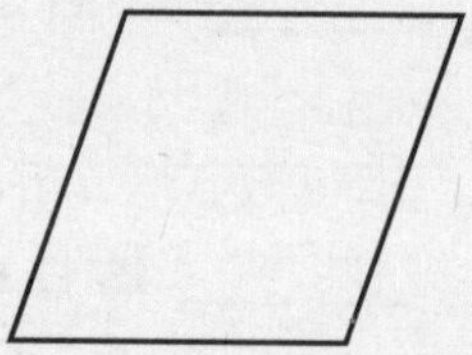

4. What is the name of this shape?

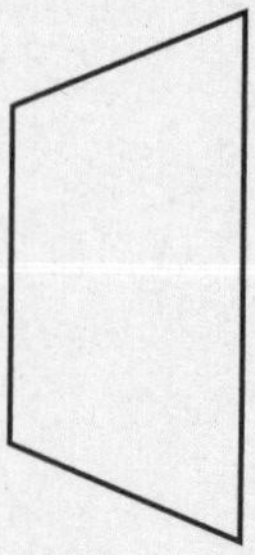

○ **A.** square

○ **B.** rectangle

○ **C.** parallelogram

○ **D.** trapezoid

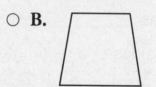

5. Which shape is **not** a parallelogram?

○ **A.** 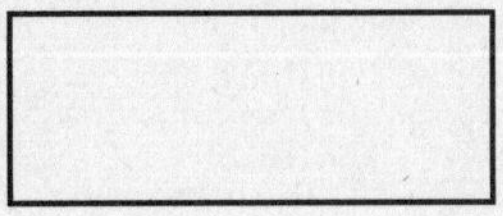

○ **B.** 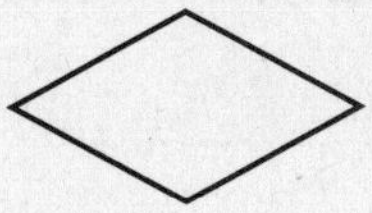

○ **C.** 

○ **D.** 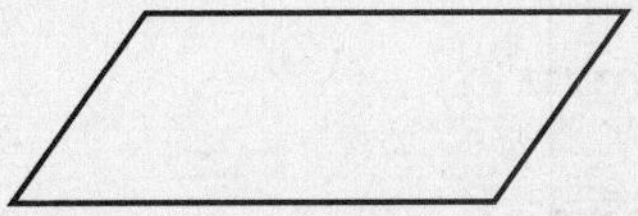

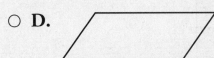

6. Which quadrilateral must have 4 equal sides?

○ **A.** rectangle

○ **B.** parallelogram

○ **C.** trapezoid

○ **D.** rhombus

7. Courtney drew a quadrilateral. The sides are not all equal in length. Which shape did Courtney draw?

○ **A.** trapezoid

○ **B.** hexagon

○ **C.** rhombus

○ **D.** square

8. Jerry made this octagon from three shapes.

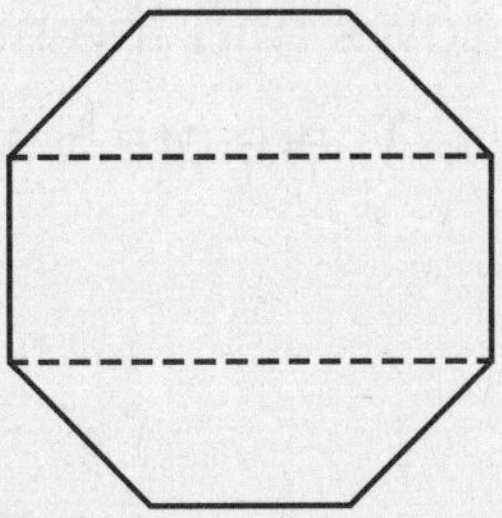

A. Draw the three shapes used to make this octagon.

B. What are the names of the shapes used to make the octagon?

____

Common Core State Standard:
3.G.2

Area of Shapes

Getting the Idea

Remember, area is the number of square units that cover a two-dimensional shape.

Example 1

What is the area of this rectangle?

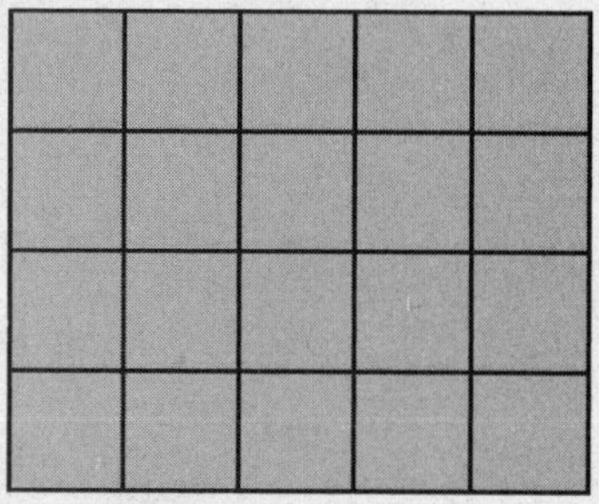

Key:  = 1 square inch

Strategy **Use multiplication.**

Step 1 Find the number of rows and the number in each row.

There are 4 rows. There are 5 squares in each row.

Step 2 Multiply.

$4 \times 5 = 20$

Step 3 Find the units.

Each square equals 1 square inch.

So, 20 squares equal 20 square inches.

Solution **The area of the rectangle is 20 square inches.**

You can also use repeated addition to find the area of a rectangle.

In Example 1 there are 4 rows with 5 squares in each row.

$5 + 5 + 5 + 5 = 20$

You can write a fraction to describe part of the area of a shape.

Remember, the numerator is the top number in a fraction.
The denominator is the bottom number in a fraction.

Example 2

Peter made a diagram of his vegetable garden. The garden is in the shape of a rectangle. There are 4 equal rows of vegetables.

Peter's Garden

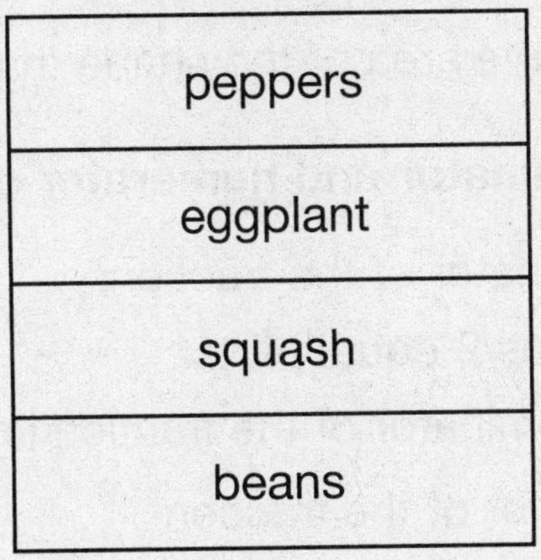

What fraction of the garden are peppers?

Strategy **Write a fraction to describe the part of the rectangle that is peppers.**

Step 1 Find the denominator of the fraction.

The total number of rows is the denominator of the fraction.

There are 4 equal rows.

The denominator of the fraction is 4.

Step 2 Find the numerator of the fraction.

There is 1 row of peppers.

The numerator of the fraction is 1.

Step 3 Write the fraction for the part of the garden that is peppers.

There is 1 row of peppers. There are 4 equal rows.

So, peppers are $\frac{1}{4}$ of the area of the whole garden.

Solution **Peppers are $\frac{1}{4}$ of the garden.**

Example 3

Carmen drew the shape below. She shaded 1 equal part.

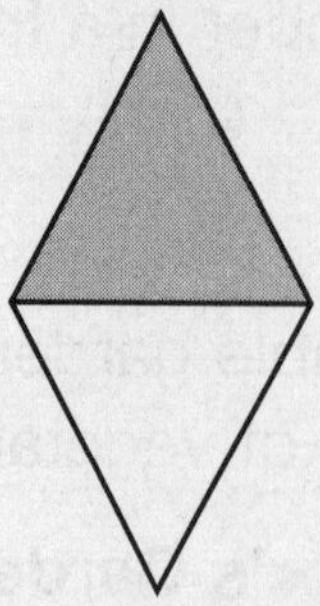

Write a fraction to describe the area of the shape that is shaded.

Strategy **Find the denominator and numerator of the fraction.**

Step 1 Find the denominator of the fraction.

The shape has 2 equal parts.

So, the denominator of the fraction is 2.

Step 2 Find the numerator of the fraction.

There is 1 shaded part.

So, the numerator of the fraction is 1.

Step 3 Write the fraction.

There is 1 shaded part. There are 2 equal parts.

So, $\frac{1}{2}$ of the shape is shaded.

Solution $\mathbf{\frac{1}{2}}$ **of the shape is shaded.**

Coached Example

Annie painted a wall in her room. She painted 2 equal parts white. She painted the other equal part blue. What fraction describes the area of the wall that is painted blue?

Use a model.

Draw a rectangle. Make 3 equal parts.

Shade one part.

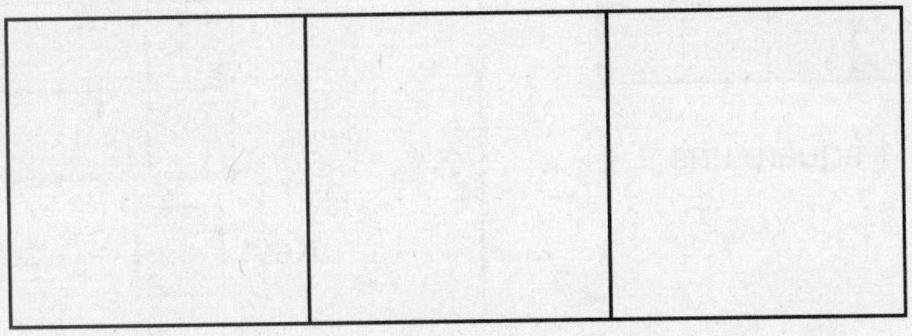

Write a fraction to describe the shaded part of the rectangle.

Find the denominator of the fraction.

How many equal parts are there? ____________

Find the numerator of the fraction.

How many shaded parts are there? ____________

Write the fraction.

What fraction names the shaded part of the rectangle? ____________

The fraction ________ describes the area of the wall that is painted blue.

Lesson Practice

Choose the correct answer.

Use the diagram below for questions 1 and 2.

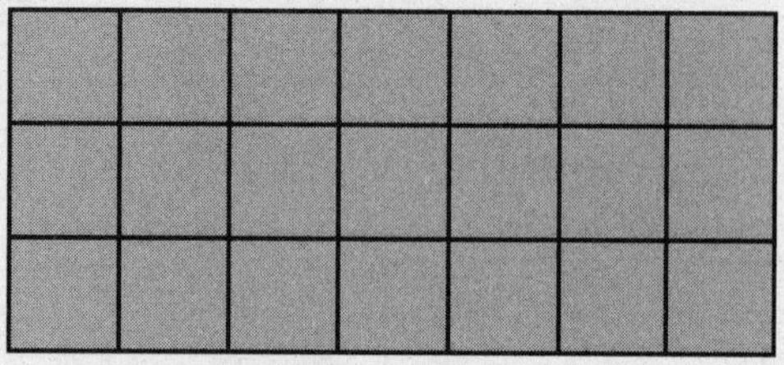

Key:  = 1 square unit

1. Which can be used to find the area of the rectangle?

   ○ **A.** 3 + 3 + 3 = □
   ○ **B.** 7 + 7 + 7 = □
   ○ **C.** 3 + 7 + 3 = □
   ○ **D.** 7 + 3 + 7 = □

2. What is the area of the rectangle?

   ○ **A.** 9 square units
   ○ **B.** 13 square units
   ○ **C.** 17 square units
   ○ **D.** 21 square units

Use the diagram below for questions 3 and 4.

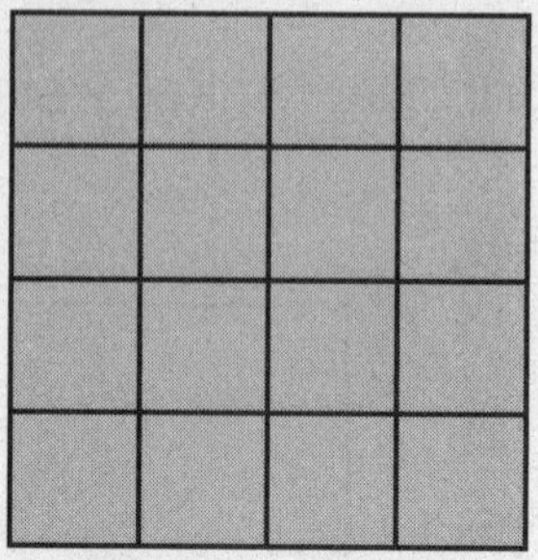

Key:  = 1 square meter

3. Which can be used to find the area of the rectangle?

   ○ **A.** 4 × 2 = □
   ○ **B.** 4 × 4 = □
   ○ **C.** 8 × 4 = □
   ○ **D.** 8 × 8 = □

4. What is the area of the rectangle?

   ○ **A.** 8 square meters
   ○ **B.** 16 square meters
   ○ **C.** 32 square meters
   ○ **D.** 64 square meters

5. What fraction describes the area of the shaded part of the hexagon?

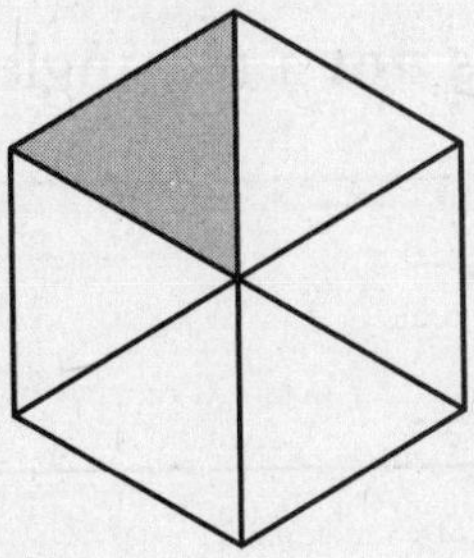

○ A. $\frac{1}{6}$

○ B. $\frac{1}{4}$

○ C. $\frac{1}{3}$

○ D. $\frac{1}{2}$

6. What fraction describes the area of the shaded part of the square?

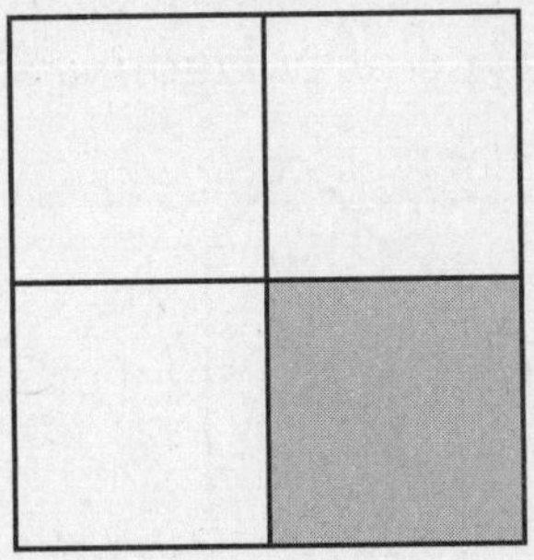

○ A. $\frac{1}{2}$

○ B. $\frac{1}{3}$

○ C. $\frac{1}{4}$

○ D. $\frac{1}{8}$

7. A diagram of a classroom floor is shown below.

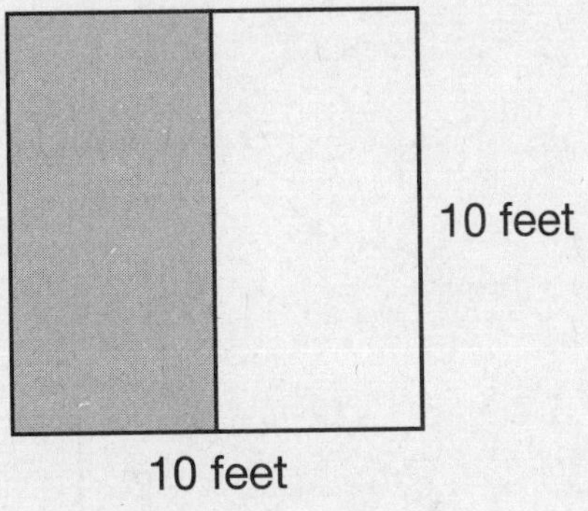

A. What is the area of the classroom floor?

____

B. The shaded part shows the area of the floor with carpet. What fraction of the classroom floor is covered by carpet?

____

Domain 5: Cumulative Assessment for Lessons 34–36

1. What type of polygon is shown?

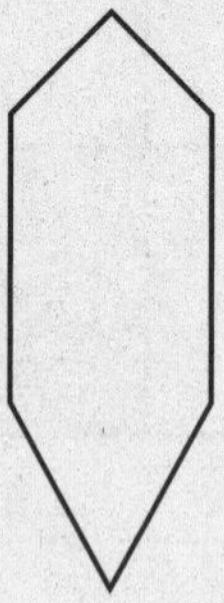

- ○ **A.** hexagon
- ○ **B.** rectangle
- ○ **C.** octagon
- ○ **D.** pentagon

2. Which two-dimensional shape has the greatest number of sides?

- ○ **A.** square
- ○ **B.** pentagon
- ○ **C.** hexagon
- ○ **D.** octagon

3. Which is **not** a rectangle?

○ **A.** 

○ **B.** 

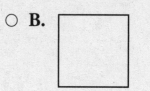

○ **C.** 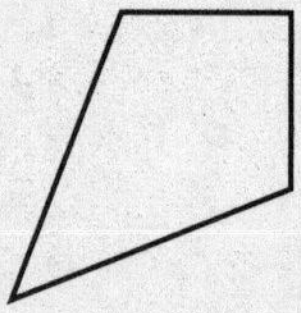

○ **D.** 

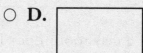

4. Which describes the area of the shaded part of the rectangle?

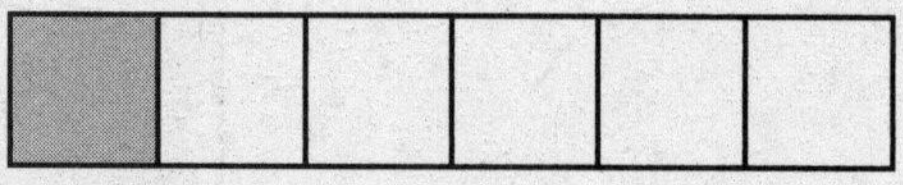

Key: 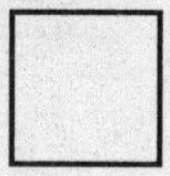 = 1 square centimeter

- ○ **A.** $\frac{1}{3}$
- ○ **B.** $\frac{1}{4}$
- ○ **C.** $\frac{1}{6}$
- ○ **D.** $\frac{1}{8}$

5. Which is **not** a quadrilateral?

- ○ **A.** square
- ○ **B.** trapezoid
- ○ **C.** parallelogram
- ○ **D.** pentagon

6. Brad drew the dotted lines on the rhombus. Which shapes did Brad make?

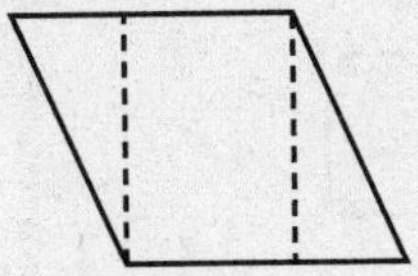

- ○ **A.** 2 triangles and 1 rectangle
- ○ **B.** 2 parallelograms and 1 triangle
- ○ **C.** 1 rectangle and 2 trapezoids
- ○ **D.** 1 square and 2 rectangles

Use the square below for questions 7 and 8.

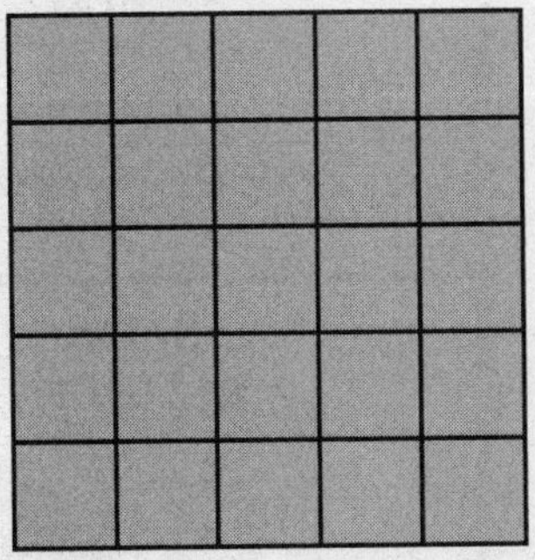

Key: ▭ = 1 square centimeter

7. Which can you use to find the area of the square?

- ○ **A.** $4 \times 5 = \square$
- ○ **B.** $4 \times 4 = \square$
- ○ **C.** $5 \times 5 = \square$
- ○ **D.** $5 \times 6 = \square$

8. What is the area of the square?

- ○ **A.** 10 square centimeters
- ○ **B.** 16 square centimeters
- ○ **C.** 20 square centimeters
- ○ **D.** 25 square centimeters

9. A diagram of a bulletin board in Ms. Peimer's classroom is shown below.

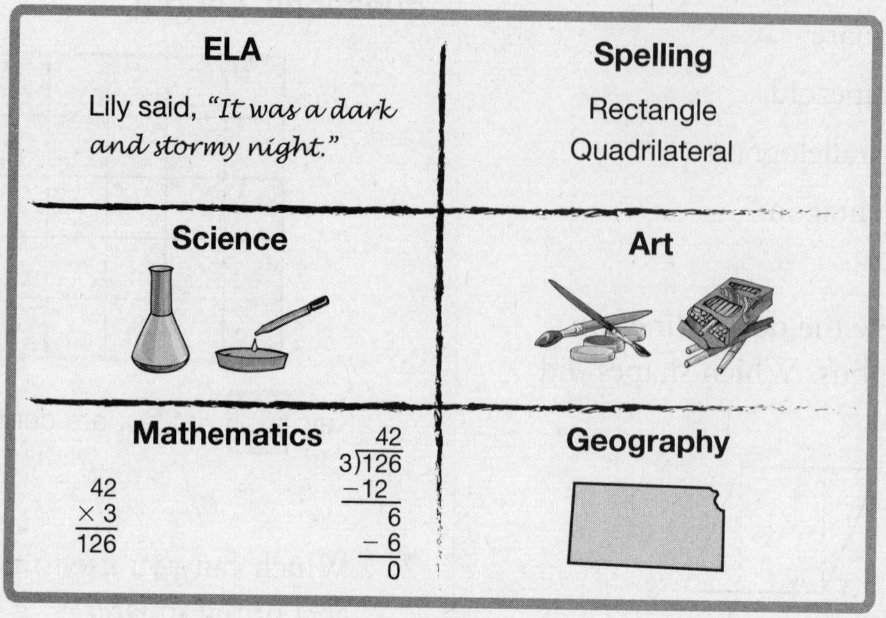

What fraction describes the area of the bulletin board titled Mathematics?

____

10. Carl drew a diagonal inside a square with 2-inch sides. He made 2 triangles that are the same shape and size. He shaded 1 part of the square.

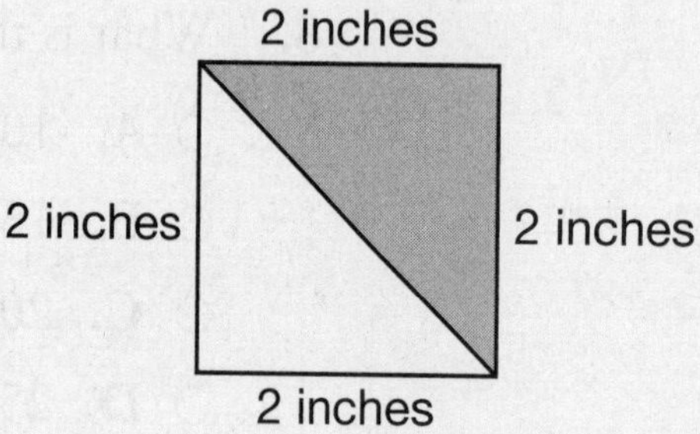

A. What 2 other names can you use to describe the square?

____

B. What fraction describes the area of the shaded part of the square? Explain your answer.

____

____

Glossary

A

add (addition) to combine two or more numbers to find the sum or total (Lesson 5)

addend numbers that are added (Lesson 5)

A.M. the hours between 12 midnight and 12 noon (Lesson 23)

angle where two sides of a polygon meet (Lesson 34)

area the amount of space that covers a figure or a region (Lesson 27)

array shows equal groups in rows and columns (Lesson 9)

associative property of addition the property that states that changing the grouping of the addends does not change the sum (Lesson 3)

associative property of multiplication the property that states that changing the grouping of the factors does not change the product (Lesson 15)

B

bar graph a graph that uses bars of different lengths to compare data (Lesson 31)

base-ten numerals a way of showing a number using each digit's place value (Lesson 1)

C

capacity the measure of how much a container can hold (Lesson 25)

centimeter (cm) a metric unit of length; 1 centimeter = 10 millimeters (Lesson 32)

circle a two-dimensional shape with a curved side (Lesson 34)

commutative property of addition the property that states that changing the order of the addends does not change the sum (Lesson 3)

commutative property of multiplication the property that states that changing the order of the factors does not change the product (Lesson 13)

D

data information (Lesson 30)

denominator the bottom part of a fraction that tells the total number of equal parts (Lesson 19)

difference the answer to a subtraction problem (Lesson 6)

digit any of the numerals 0, 1, 2, 3, 4, 5, 6, 7, 8, and 9 (Lesson 1)

distributive property of multiplication the property that states that multiplying a sum by a factor is the same as multiplying each addend by the factor and adding the products (Lessons 13, 28)

dividend a number being divided (Lesson 16)

division an operation that separates objects into equal groups (Lesson 16)

divisor a number that a dividend is divided by (Lesson 16)

E

elapsed time the amount of time from the start to the finish of an event (Lesson 23)

equivalent fractions two or more fractions that name the same value but have different numerators and denominators (Lesson 21)

estimate a number close to the exact answer (Lesson 8)

even number a number with 0, 2, 4, 6, or 8 in the ones place (Lesson 4)

expanded form a way of writing numbers showing the value of each digit (Lesson 1)

F

fact family a set of related facts that use the same numbers (Lesson 16)

factor a number being multiplied in a multiplication problem (Lesson 9)

fraction a number that names part of a whole (Lesson 19)

G

gram a unit of mass; 1,000 grams = 1 kilogram (Lesson 24)

H

hexagon a polygon with 6 sides and 6 angles (Lesson 34)

hour a unit of time; 60 minutes = 1 hour (Lesson 23)

I

identity property of addition the property that states that the sum of any addend and 0 is the addend (Lesson 3)

inch (in.) a customary unit of length used to measure small objects (Lesson 32)

inverse operations operations that undo each other; addition is the inverse operation of subtraction; multiplication is the inverse operation of division (Lesson 16)

is equal to (=) a symbol used to compare two numbers to show that they have the same value (Lessons 2, 22)

is greater than (>) a symbol used to compare two numbers, with the greater number given first (Lessons 2, 22)

is less than (<) a symbol used to compare two numbers, with the lesser number given first (Lessons 2, 22)

K

kilogram a unit of mass; 1 kilogram = 1,000 grams (Lesson 24)

L

length is the measure of how long, wide, or tall something is (Lesson 32)

line plot a graph that uses Xs or dots above a number line to record data (Lesson 33)

liter a metric unit of capacity; 1 liter = 1,000 milliliters (Lesson 25)

M

mass a measure of how much matter an object has (Lesson 24)

milliliter a metric unit of capacity; 1,000 milliliters = 1 liter (Lesson 25)

millimeter (mm) a metric unit of length; 10 millimeters = 1 centimeter (Lesson 32)

minuend the number that is subtracted from in a subtraction problem (Lesson 6)

minute a unit of time; 60 seconds = 1 minute (Lesson 23)

multiple a product of two numbers (Lessons 6, 14)

multiplication an operation that joins equal groups (Lesson 9)

N

number name a way of showing a number using words (Lesson 1)

number pattern a group of numbers that follows a rule (Lesson 4)

numerator the top part of a fraction that tells the number of equal parts being considered (Lesson 19)

O

octagon a polygon with 8 sides and 8 angles (Lesson 34)

odd number a number with 1, 3, 5, 7, or 9 in the ones place (Lesson 4)

P

parallel remain the same distance apart; never meet (Lesson 35)

parallelogram a quadrilateral with 2 pairs of opposite sides that are parallel and with opposite sides the same length (Lesson 35)

pentagon a polygon with 5 sides and 5 angles (Lesson 34)

perimeter the distance around a figure or an object (Lesson 26)

picture graph a graph that uses pictures or symbols to display and compare data (Lesson 30)

place value the value of each digit in a number (Lesson 1)

place-value chart a table that shows the value of each digit in a number (Lesson 1)

place-value system a number system in which the position of a digit in a number determines its value (Lesson 1)

P.M. the hours between 12 noon and 12 midnight (Lesson 23)

polygon a two-dimensional shape with straight sides (Lesson 34)

product the answer to a multiplication problem (Lesson 9)

Q

quadrilateral a polygon with 4 sides and 4 angles (Lessons 34, 35)

quotient the answer to a division problem (Lesson 16)

R

rectangle a quadrilateral with 2 pairs of opposite sides that are parallel, with opposite sides the same length, and with 4 square corners (Lesson 35)

regroup to exchange amounts of equal value to rename a number (Lessons 5, 6)

repeated addition adding the same number multiple times; has the same result as multiplication (Lesson 9)

repeated subtraction subtracting the same number multiple times; is related to division (Lesson 16)

round a way to estimate numbers to a specific place value, like tens or hundreds (Lesson 7)

rounding rules look at the digit to the right of the place you are rounding to. If the digit is less than 5, round down. If the digit is 5 or greater, round up (Lesson 7)

rhombus a quadrilateral with 4 equal sides and 2 pairs of opposite parallel sides (Lesson 35)

S

scale a part of a bar graph that tells the value of each bar (Lesson 31)

scale a tool used to measure mass or how heavy an object is (Lesson 24)

side a straight edge in a two-dimensional shape (Lesson 34)

square a rectangle with 4 equal sides and 4 square corners (Lesson 35)

square unit a square that has a side length of 1 unit (Lesson 27)

subtract (subtraction) a way of taking one number away from another to find the difference (Lesson 6)

subtrahend the number that is subtracted in a subtraction problem (Lesson 6)

sum the answer to an addition problem (Lesson 5)

T

trapezoid a quadrilateral with exactly 1 pair of parallel sides (Lesson 35)

triangle a polygon with 3 sides and 3 angles (Lesson 34)

two-dimensional shape a flat shape that has length and width (Lesson 34)

U

unit fraction a fraction with 1 as the numerator (Lesson 19)

W

whole numbers any of the numbers 0, 1, 2, 3, and so on (Lesson 1)

Crosswalk Coach for the Common Core State Standards
Mathematics, Grade 3

Summative Assessment: Domains 1–5

Name: ______________________________

Session 1

1. A movie theater has 518 seats. To the nearest hundred, about how many seats are in the theater?

○ **A.** 600

○ **B.** 520

○ **C.** 510

○ **D.** 500

2. Don made this number pattern.

43 51 59 ? 75 83

What is the missing number in Don's pattern?

○ **A.** 71

○ **B.** 67

○ **C.** 63

○ **D.** 61

3. Melanie has collected 148 stamps from Spain and 153 stamps from France. Her goal is to collect 400 stamps from both countries. About how many stamps does Melanie need to collect to reach her goal?

○ **A.** 10

○ **B.** 50

○ **C.** 100

○ **D.** 150

4. Which shows $\frac{2}{3}$ of the square shaded?

○ **A.**

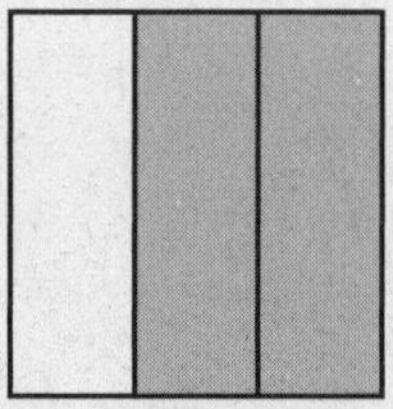

○ **B.**

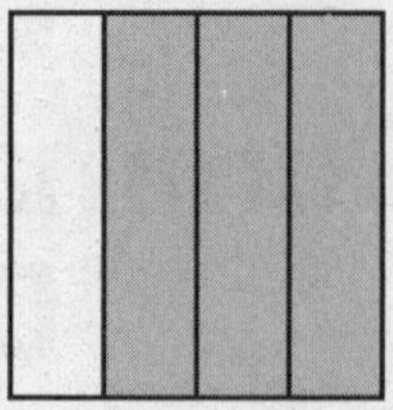

○ **C.**

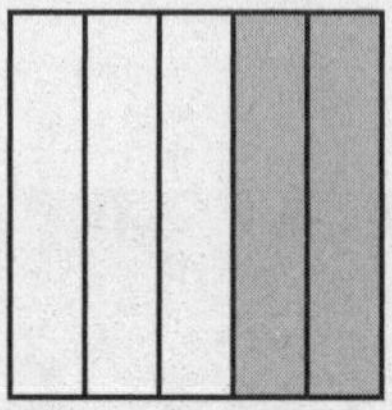

○ **D.**

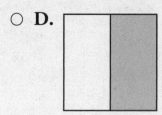

5. Kevin made some muffins. It took him 35 minutes to mix the ingredients and 25 minutes to bake the muffins. How much time in all did it take Kevin to make the muffins?

○ **A.** 10 minutes

○ **B.** 50 minutes

○ **C.** 55 minutes

○ **D.** 60 minutes

6. Which clock shows 8:12?

○ **A.**

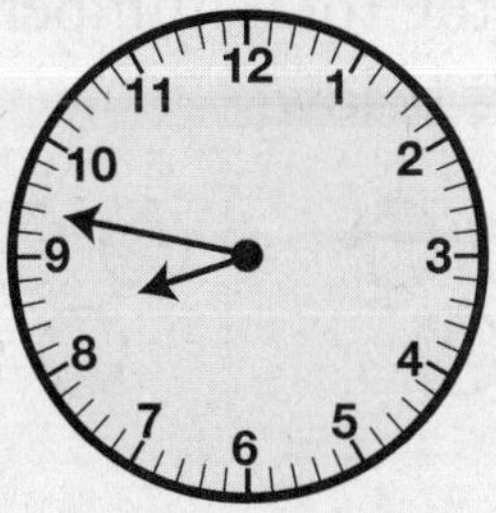

○ **B.**

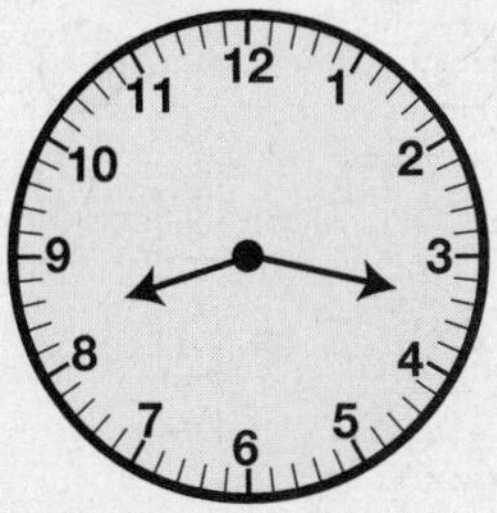

○ **C.**

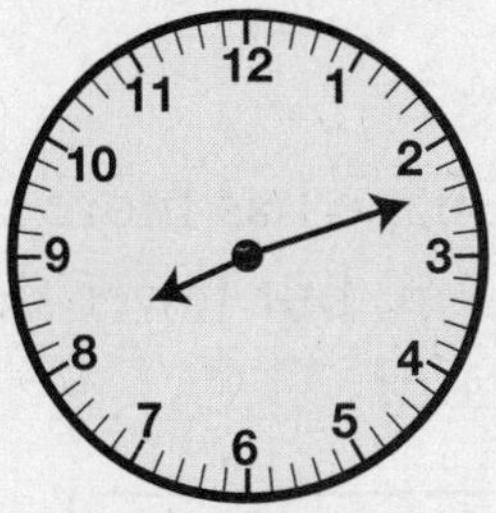

○ **D.**

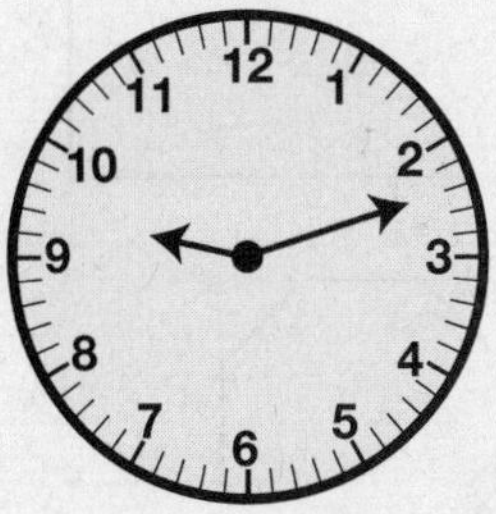

7. The table shows the number of hours each class spent helping at an animal shelter last year.

Hours at Animal Shelter

Class	Number of Hours
Grade 3	125
Grade 4	207
Grade 5	173
Grade 6	182

How many hours did Grade 4 and Grade 5 spend at the shelter last year?

○ **A.** 280 hours

○ **B.** 370 hours

○ **C.** 380 hours

○ **D.** 389 hours

8. Kayla received $200 for winning a writing contest. She bought an MP3 player for $59 and donated $35 to charity. She saved the rest in the bank. How much did Kayla save in the bank?

○ **A.** $96

○ **B.** $106

○ **C.** $116

○ **D.** $126

9. What is the perimeter of the square?

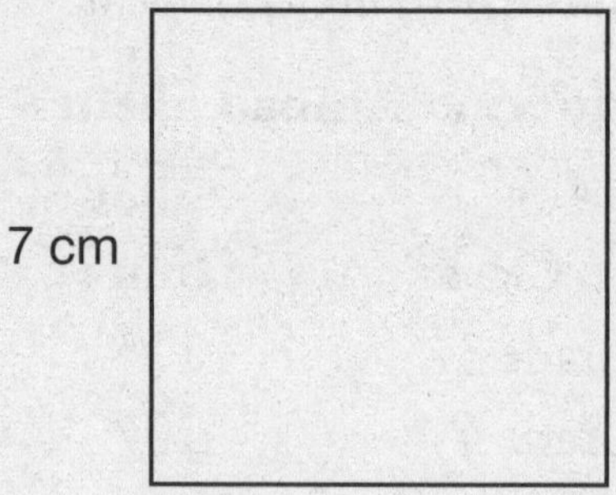

○ **A.** 49 centimeters

○ **B.** 28 centimeters

○ **C.** 22 centimeters

○ **D.** 14 centimeters

10. A company sold 628 copies of the Sunday paper and 382 copies of the Monday paper. How many more Sunday papers than Monday papers did the company sell?

○ **A.** 246

○ **B.** 256

○ **C.** 366

○ **D.** 300

11. Which sentence can be used to find the total number of counters below?

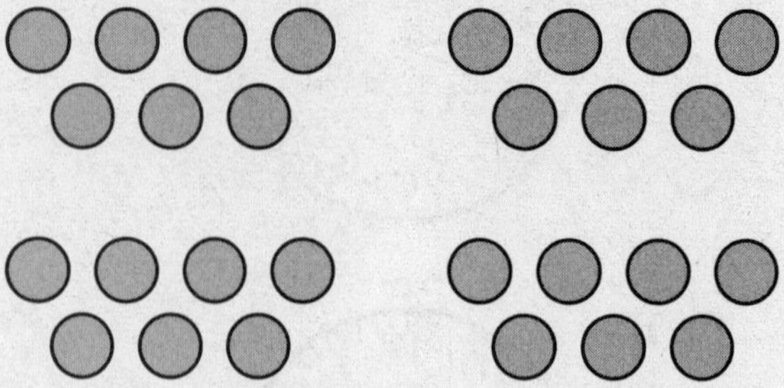

○ **A.** $4 \times 7 = \square$

○ **B.** $4 + 7 = \square$

○ **C.** $7 \times 7 = \square$

○ **D.** $6 + 7 = \square$

12. Which shape has more sides than a pentagon, but fewer sides than an octagon?

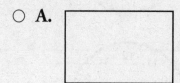

○ **A.** 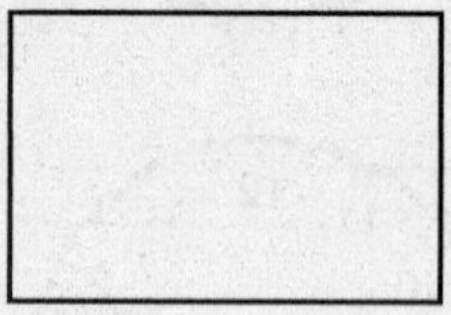

○ **B.** 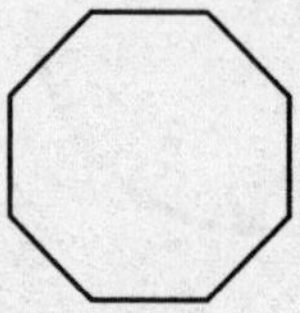

○ **C.** 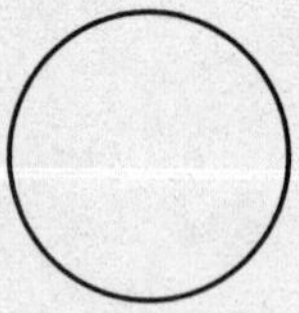

○ **D.** 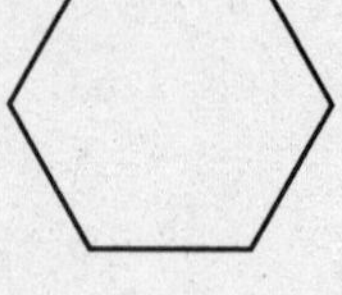

13. Which number makes both sentences true?

$7 \times \square = 49$

$49 \div \square = 7$

○ **A.** 6
○ **B.** 7
○ **C.** 8
○ **D.** 9

14. There are 12 eggs in a dozen. Claude has 7 dozen eggs. How many eggs does he have?

○ **A.** 80
○ **B.** 84
○ **C.** 86
○ **D.** 96

15. Which is the same as $\frac{6}{6}$?

○ **A.** 1
○ **B.** 2
○ **C.** 5
○ **D.** 6

16. Carina has 12 glasses to put on 3 shelves. She wants to put the same number of glasses on each shelf.

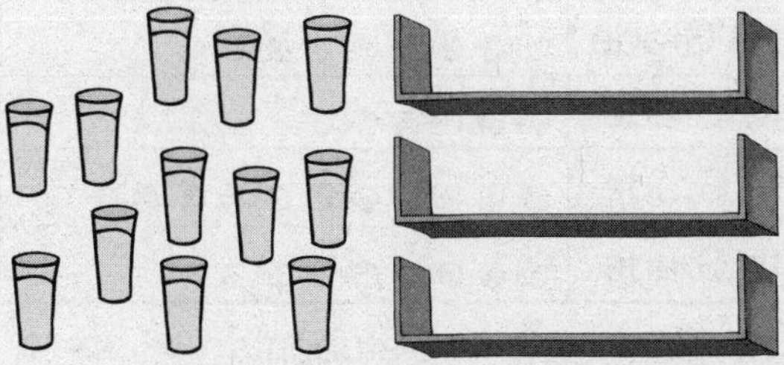

How many glasses will she put on each shelf?

○ **A.** 4
○ **B.** 5
○ **C.** 8
○ **D.** 15

17. What fraction of the triangle is shaded?

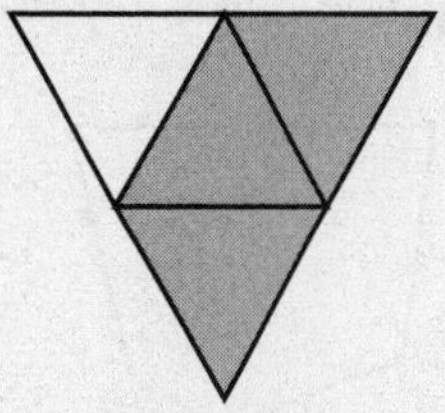

○ **A.** $\frac{1}{4}$
○ **B.** $\frac{3}{4}$
○ **C.** $\frac{4}{4}$
○ **D.** $\frac{3}{3}$

18. Justin made a picture graph about students who ride their bikes to school.

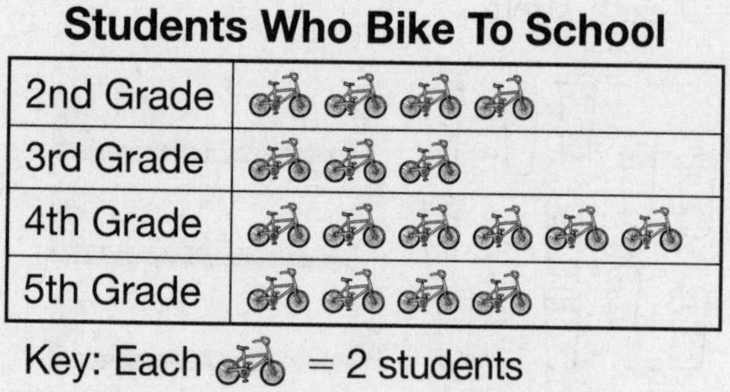

Students Who Bike To School
2nd Grade
3rd Grade
4th Grade
5th Grade
Key: Each = 2 students

How many fewer 3rd graders ride their bikes to school than 4th graders?

- ○ **A.** 3
- ○ **B.** 6
- ○ **C.** 9
- ○ **D.** 12

19. Which is the best name for this shape?

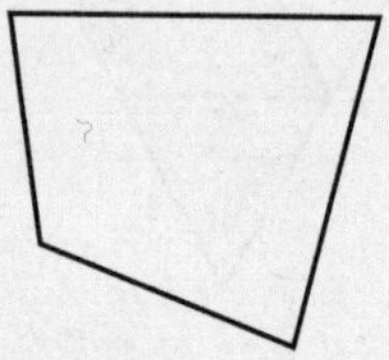

- ○ **A.** square
- ○ **B.** rhombus
- ○ **C.** rectangle
- ○ **D.** quadrilateral

20. This table shows the total number of pencils for different numbers of packages.

Pencils in Packages

Number of Packages	Number of Pencils
2	12
4	24
6	36
8	48

How many pencils are in 7 packages?

- ○ **A.** 38
- ○ **B.** 40
- ○ **C.** 42
- ○ **D.** 45

21. Which fraction is equivalent to $\frac{1}{4}$?

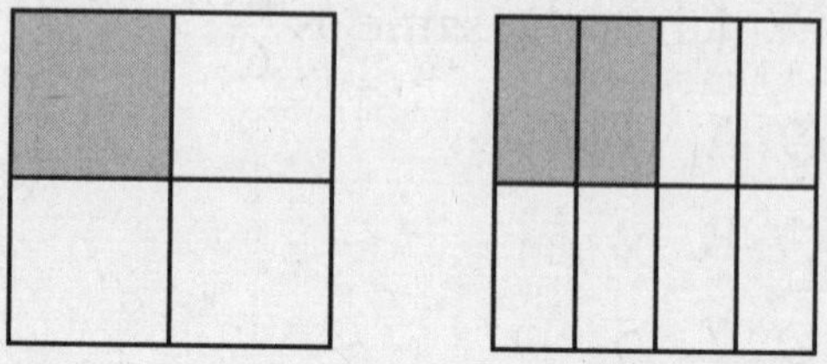

- ○ **A.** $\frac{1}{8}$
- ○ **B.** $\frac{2}{8}$
- ○ **C.** $\frac{3}{8}$
- ○ **D.** $\frac{3}{4}$

22. What number belongs in the $\square$?

$7 \times 8 = \square \times 7$

○ **A.** 1
○ **B.** 7
○ **C.** 8
○ **D.** 15

23. Which is the best estimate for the mass of these books?

○ **A.** 4 grams
○ **B.** 4 kilograms
○ **C.** 40 grams
○ **D.** 40 kilograms

24. What is the area of the shaded rectangle?

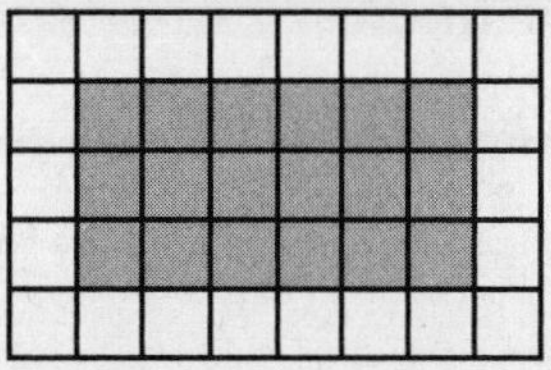

Key: ▒ = 1 square inch

○ **A.** 15 square inches
○ **B.** 18 square inches
○ **C.** 24 square inches
○ **D.** 30 square inches

25. Olivia packed 50 candles in each box. She packed 7 boxes. How many candles in all did Olivia pack?

○ **A.** 35
○ **B.** 120
○ **C.** 350
○ **D.** 570

26. Which equivalent fractions does the point show on the number line?

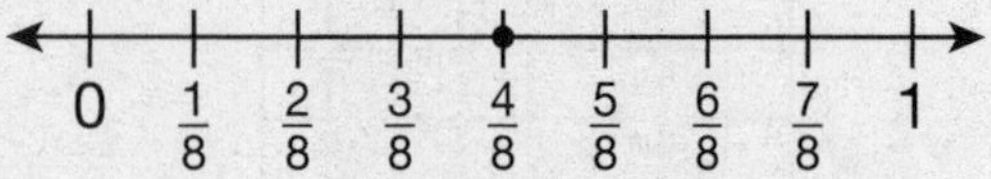

○ **A.** $\frac{4}{8}$ and $\frac{1}{4}$

○ **B.** $\frac{4}{8}$ and $\frac{1}{2}$

○ **C.** $\frac{4}{8}$ and $\frac{1}{6}$

○ **D.** $\frac{4}{8}$ and $\frac{1}{3}$

27. Which symbol makes the sentence true?

$\frac{1}{3} \bigcirc \frac{2}{3}$

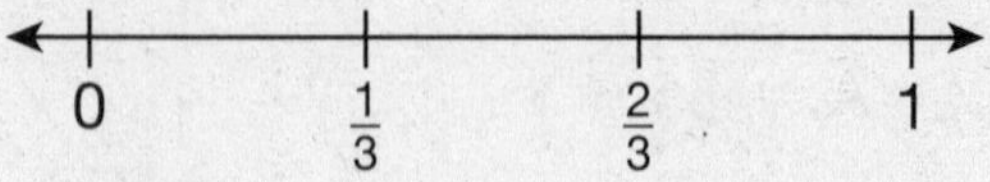

○ **A.** =

○ **B.** +

○ **C.** >

○ **D.** <

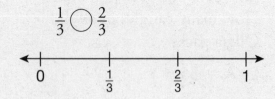

28. Eddie got a new fish tank for his birthday. He cannot carry the fish tank by himself.

Which is the best estimate of the amount of water Eddie needs to fill the fish tank?

○ **A.** 3 milliliters

○ **B.** 3 liters

○ **C.** 30 milliliters

○ **D.** 30 liters

29. This triangle has a perimeter of 22 meters.

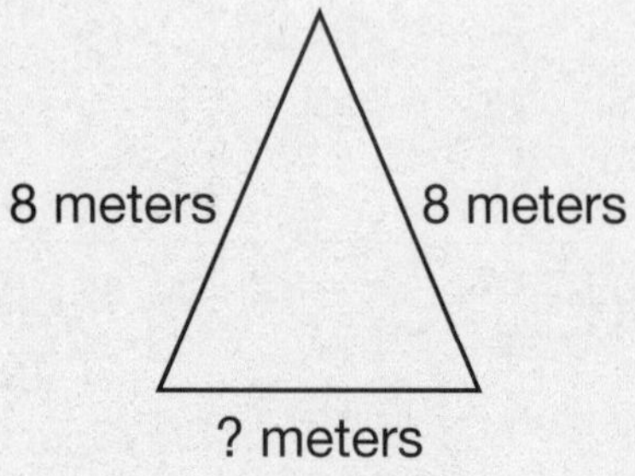

What is the missing side length?

○ **A.** 2 meters

○ **B.** 4 meters

○ **C.** 6 meters

○ **D.** 8 meters

30. Dara's class went to the school library. The bar graph shows the books that they borrowed.

Which picture graph correctly shows the data in the bar graph?

○ **A.**

○ **C.**

○ **B.**

○ **D.**

31. Which symbol makes this sentence true?

$\frac{5}{8} \bigcirc \frac{5}{6}$

- ○ **A.** >
- ○ **B.** <
- ○ **C.** =
- ○ **D.** +

32. A part of this circle is shaded.

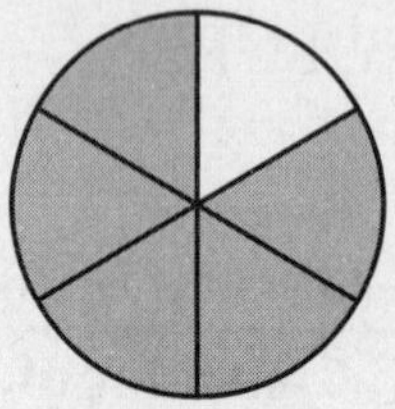

Which number line has point *R* on the fraction that shows the shaded part of the circle?

○ **A.**

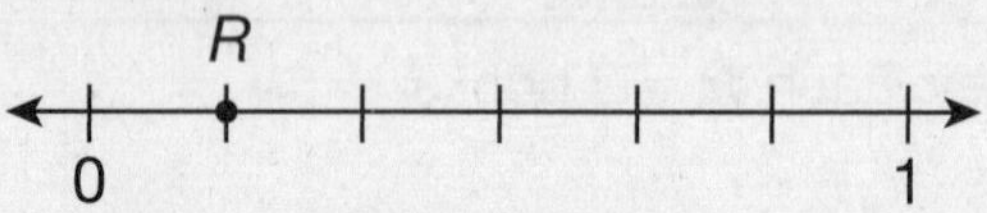

○ **B.**

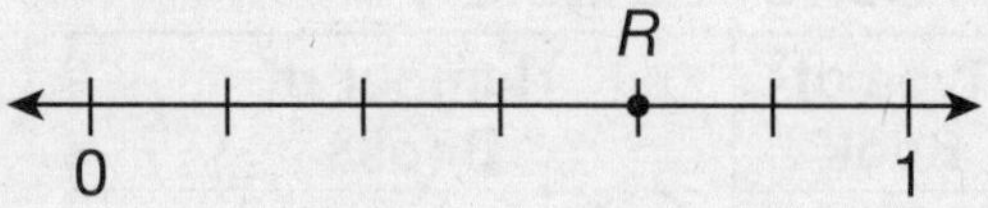

○ **C.**

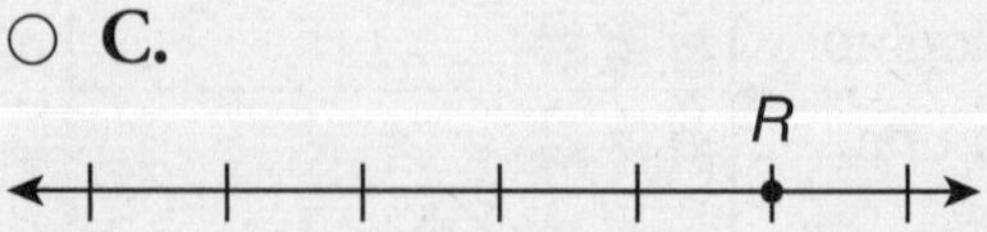

○ **D.**

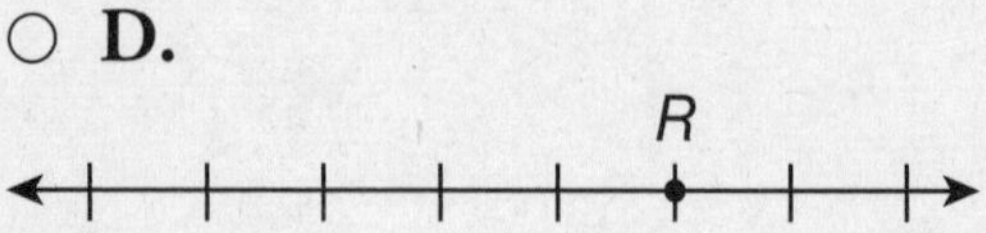

33. Dominic drew a rectangle on the board.

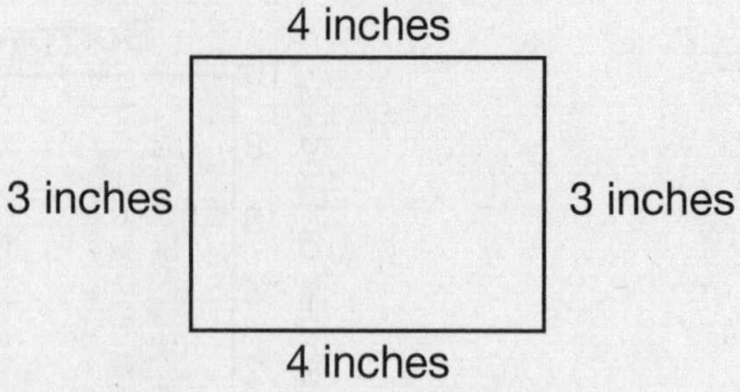

Which shows the area of Dominic's rectangle?

- ○ **A.** $4 \times 3 = \text{Area}$
- ○ **B.** $4 \times 4 = \text{Area}$
- ○ **C.** $4 + 4 + 3 + 3 = \text{Area}$
- ○ **D.** $3 + 3 + 3 + 3 = \text{Area}$

34. A napkin has a mass of 4 grams. What is the total mass of 9 napkins?

- ○ **A.** 45 grams
- ○ **B.** 36 grams
- ○ **C.** 30 grams
- ○ **D.** 13 grams

35. Multiply.

$4 \times 9 \times 2 = \square$

- ○ **A.** 8
- ○ **B.** 32
- ○ **C.** 36
- ○ **D.** 72

36. Which shape is **not** a parallelogram?

○ **A.** 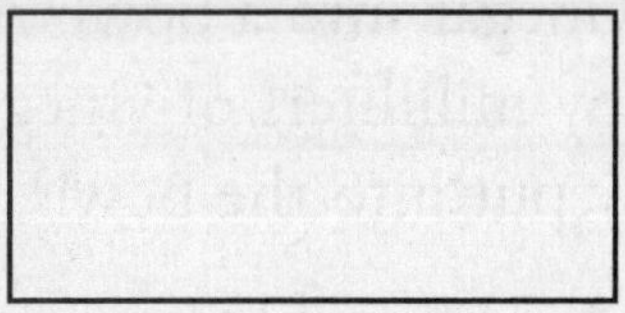

○ **B.** 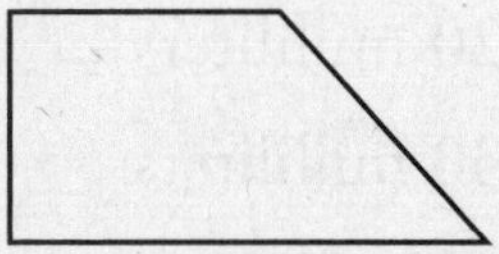

○ **C.** 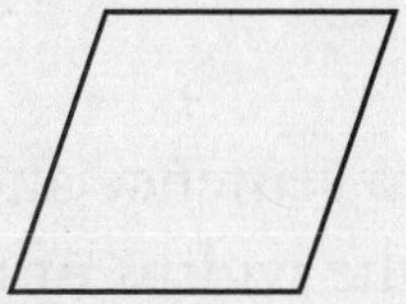

○ **D.** 

37. Which is **not** equal to 1?

○ **A.** $\frac{2}{2}$

○ **B.** $\frac{3}{3}$

○ **C.** $\frac{4}{4}$

○ **D.** $\frac{8}{1}$

38. Which sentence can be used to find the area of this figure?

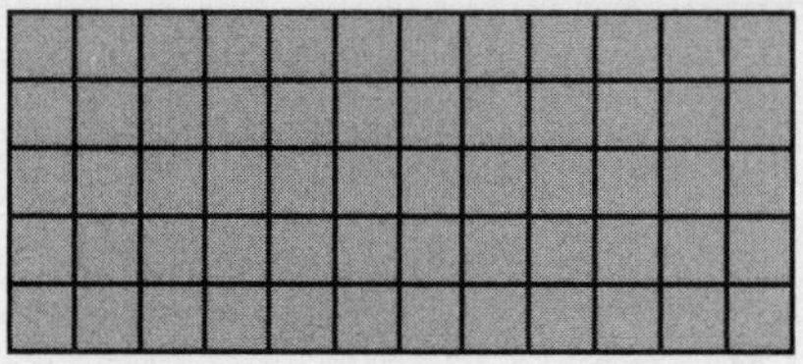

○ **A.** $(5 \times 10) + (5 \times 2)$ $=$ Area

○ **B.** $(5 + 10) \times (5 + 2)$ $=$ Area

○ **C.** $(5 + 10) + (5 + 2)$ $=$ Area

○ **D.** $(5 \times 5) + (10 \times 2)$ $=$ Area

39. Raven cut out 15 red stars and 5 blue stars for an art project. She will put 5 stars on each piece of construction paper. How many pieces of construction paper will Raven use?

○ **A.** 1

○ **B.** 3

○ **C.** 4

○ **D.** 5

40. Which fraction belongs on the line to make the sentence true?

$\frac{1}{4} >$ _____

○ **A.** $\frac{1}{2}$

○ **B.** $\frac{1}{3}$

○ **C.** $\frac{1}{4}$

○ **D.** $\frac{1}{6}$

41. Theresa crocheted rectangles to make a baby blanket. She made equal rectangles of different colors. The diagram below shows the colors she used.

White	Blue	Green	Pink
Red	Yellow	Orange	Purple

What fraction describes the area of the green part of the blanket?

○ **A.** $\frac{1}{3}$

○ **B.** $\frac{1}{4}$

○ **C.** $\frac{1}{6}$

○ **D.** $\frac{1}{8}$

42. A teaspoon can hold about 5 milliliters. Joey put 6 teaspoons of vinegar into a bowl. About how many milliliters of vinegar did Joey put into the bowl?

○ **A.** 11 milliliters

○ **B.** 20 milliliters

○ **C.** 30 milliliters

○ **D.** 40 milliliters

43. Ben wrote a sentence and measured the widths of the letters. He made the line plot below.

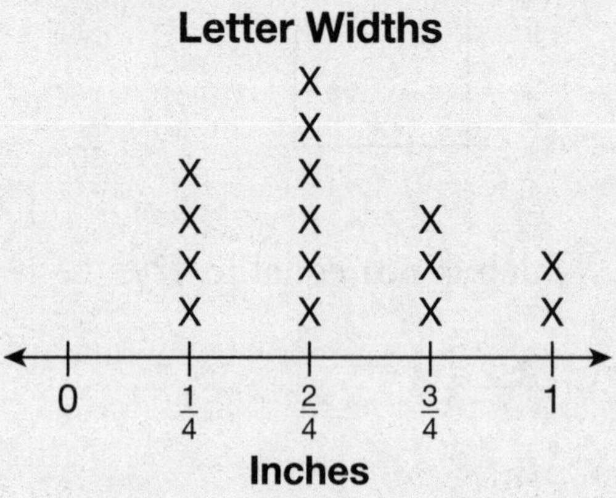

How many more $\frac{2}{4}$-inch letters are there than 1-inch letters?

○ **A.** 2

○ **B.** 4

○ **C.** 6

○ **D.** 8

STOP

Session 2

44. What fraction describes where point N is located on the number line?

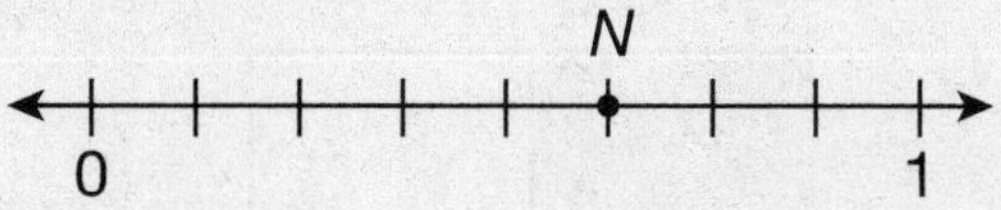

45. Mr. Robinson bought tickets to see a movie. He bought 1 adult ticket and 3 child tickets. The adult ticket cost $10. Each child ticket cost $7. How much did Mr. Robinson pay in all for the tickets?

46. The table shows the numbers of different animals in a park.

Animals at the Park

Type of Animal	Number of Animals
Squirrels	157
Hawks	96
Deer	184
Rabbits	205

How many more rabbits than hawks are in the park?

47. Kara picked 30 oranges. She put 5 oranges in each bag. How many bags of oranges did she make?

48. Look at the models below.

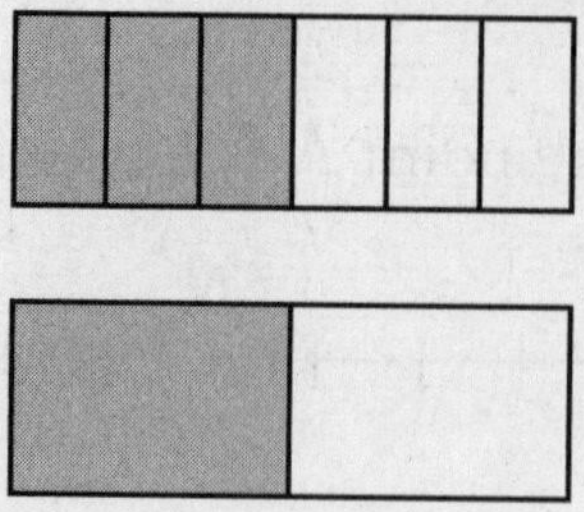

A. Write two equivalent fractions for the models.

________________ and ________________

B. Show the two equivalent fractions on the number lines below.

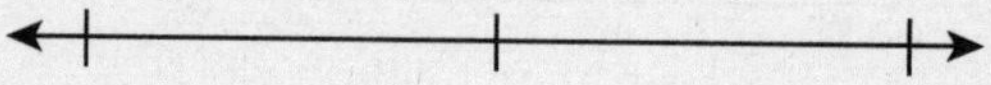

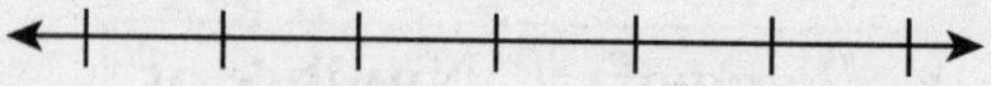

49. Shari has a poster with an area of 24 square inches.

A. On the grid below, draw two different rectangles that have an area of 24 square inches.

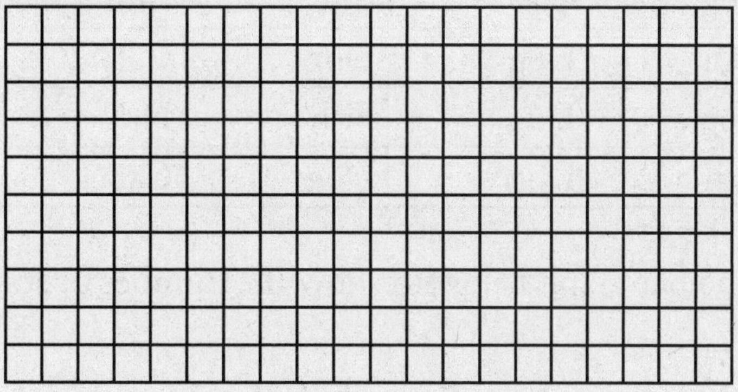

Key: □ = 1 square inch

B. What are the length and width of each rectangle that you drew for Part A?

What is the perimeter of each rectangle?

____

____

____

50. Mr. Rosner's students voted for their favorite pets. All the students wrote their favorite pets in the table below.

Favorite Pet

Dog	Cat	Cat	Bird
Fish	Dog	Dog	Cat
Dog	Dog	Fish	Fish
Fish	Dog	Dog	Cat

A. Complete the bar graph below to show the number of students who voted for each pet.

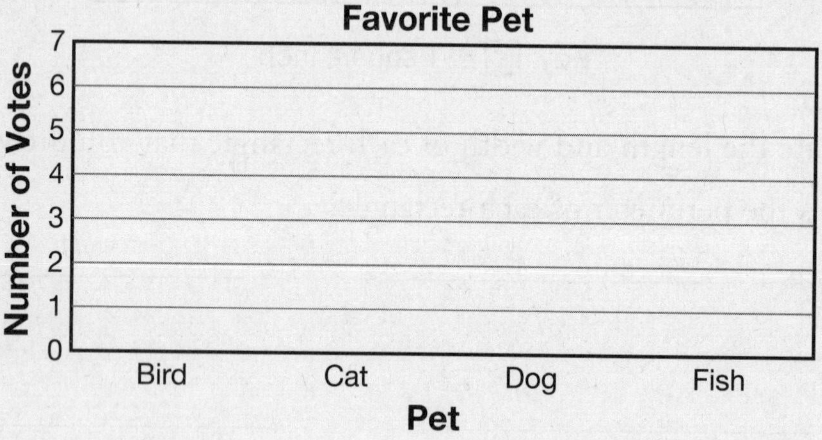

B. Based on the data in the graph, how many students' favorite pet is **not** fish? Show your work.

____

STOP

Math Tools

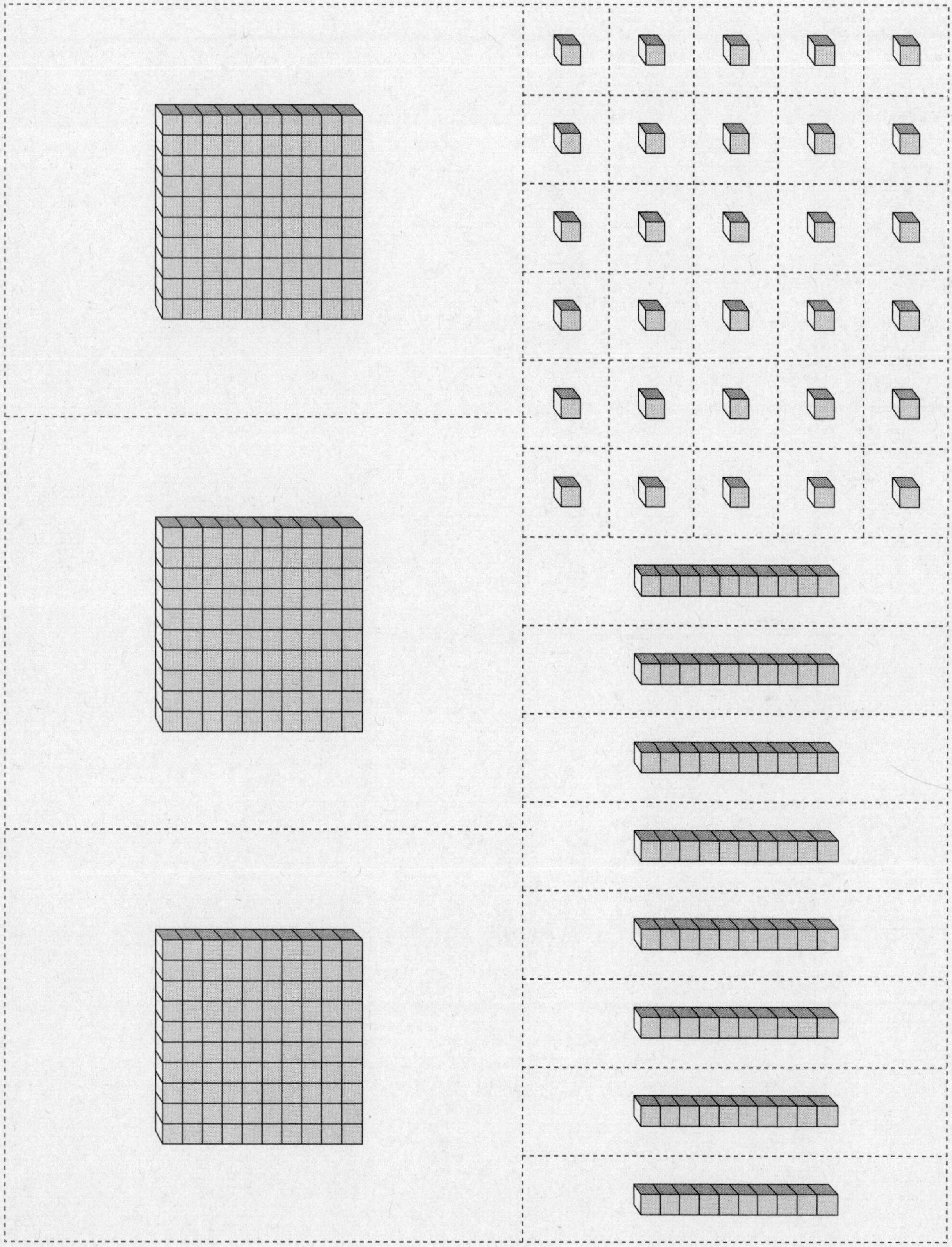

Math Tools

 Ones	
 Tens	
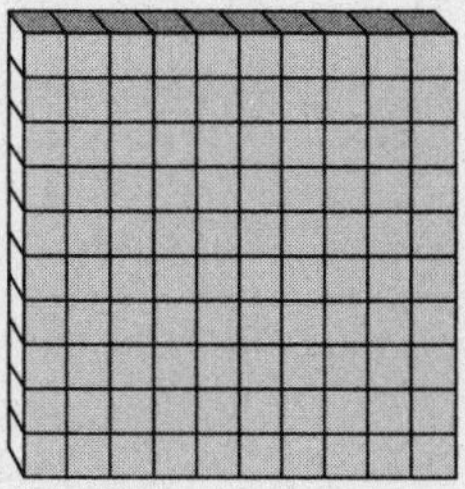 Hundreds	

Math Tools

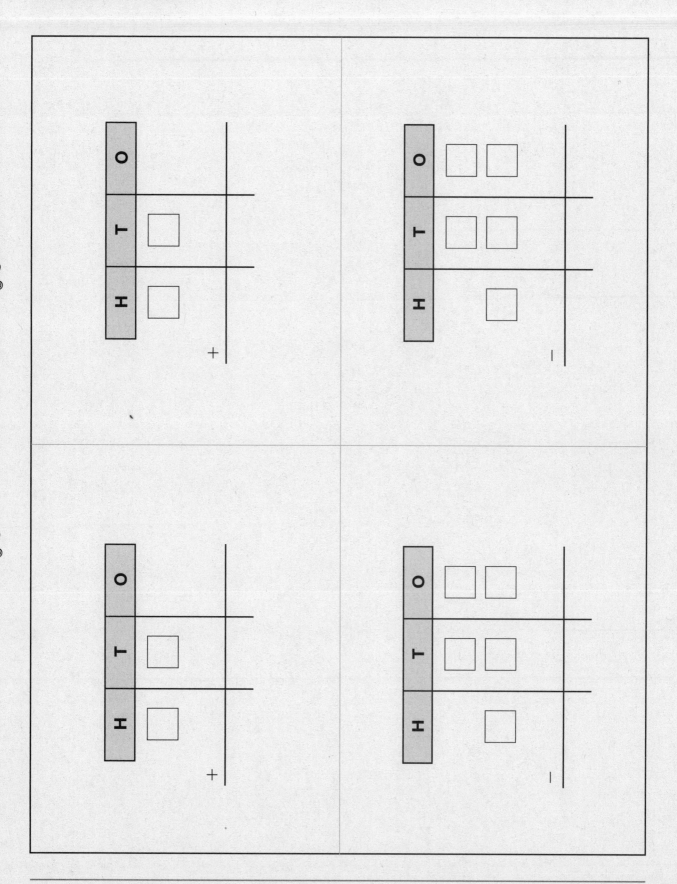

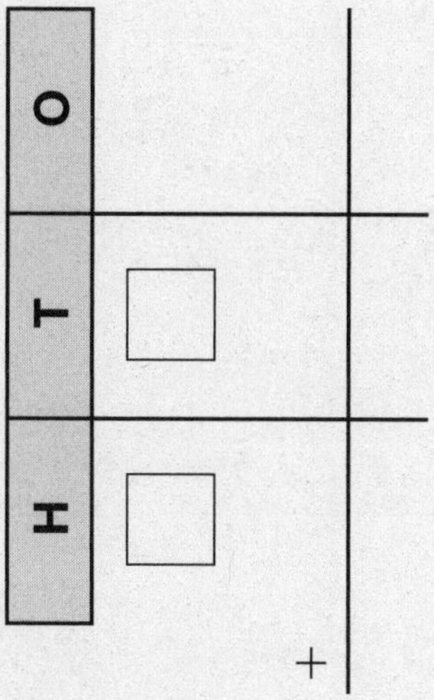

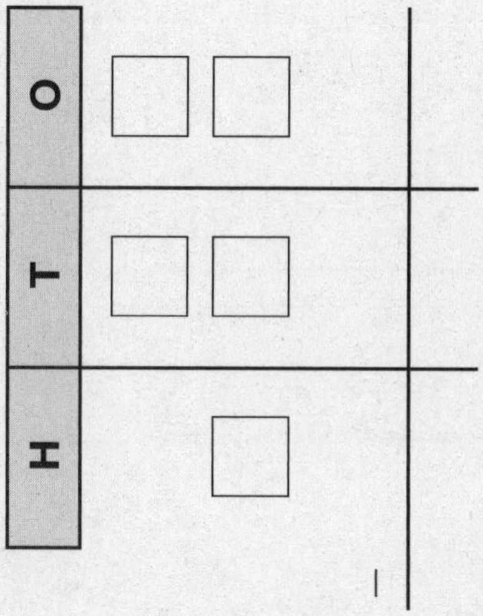

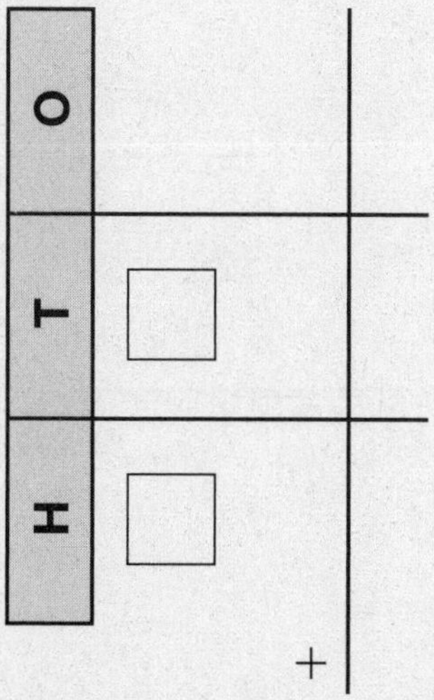

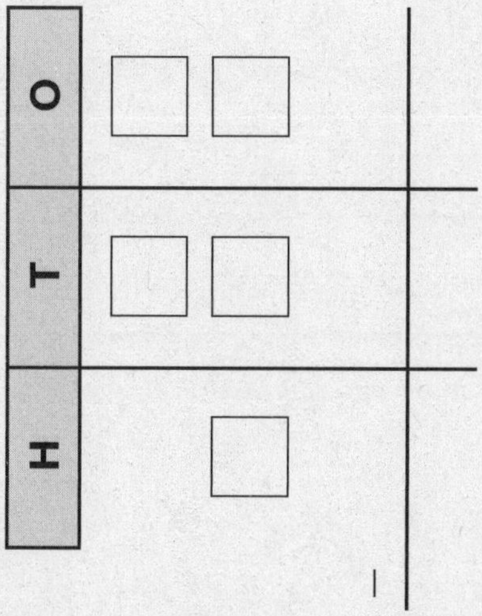

Math Tools

Hundred Thousands	Ten Thousands	Thousands	Hundreds	Tens	Ones

Hundred Thousands	Ten Thousands	Thousands	Hundreds	Tens	Ones

Math Tools

Math Tools

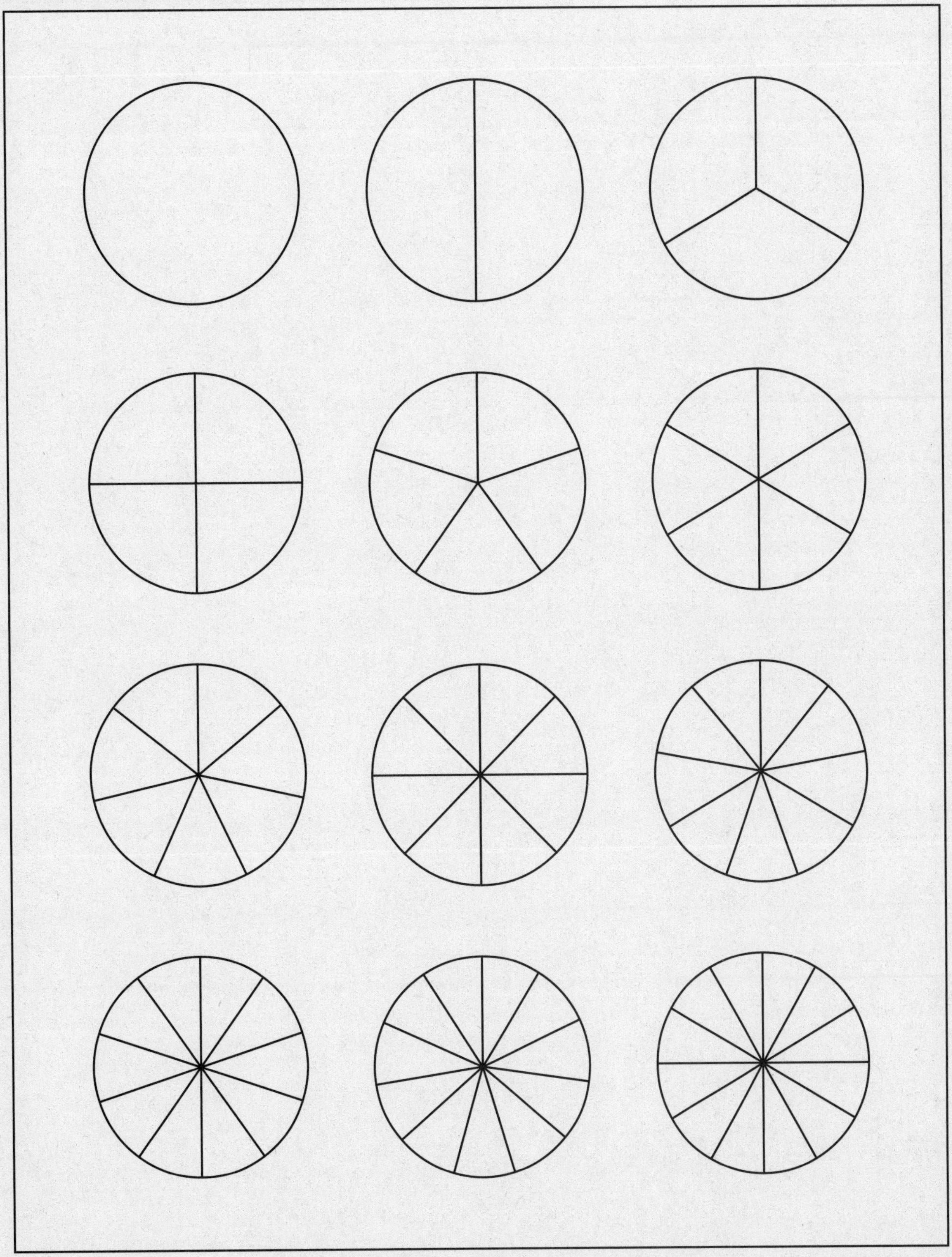

Math Tools

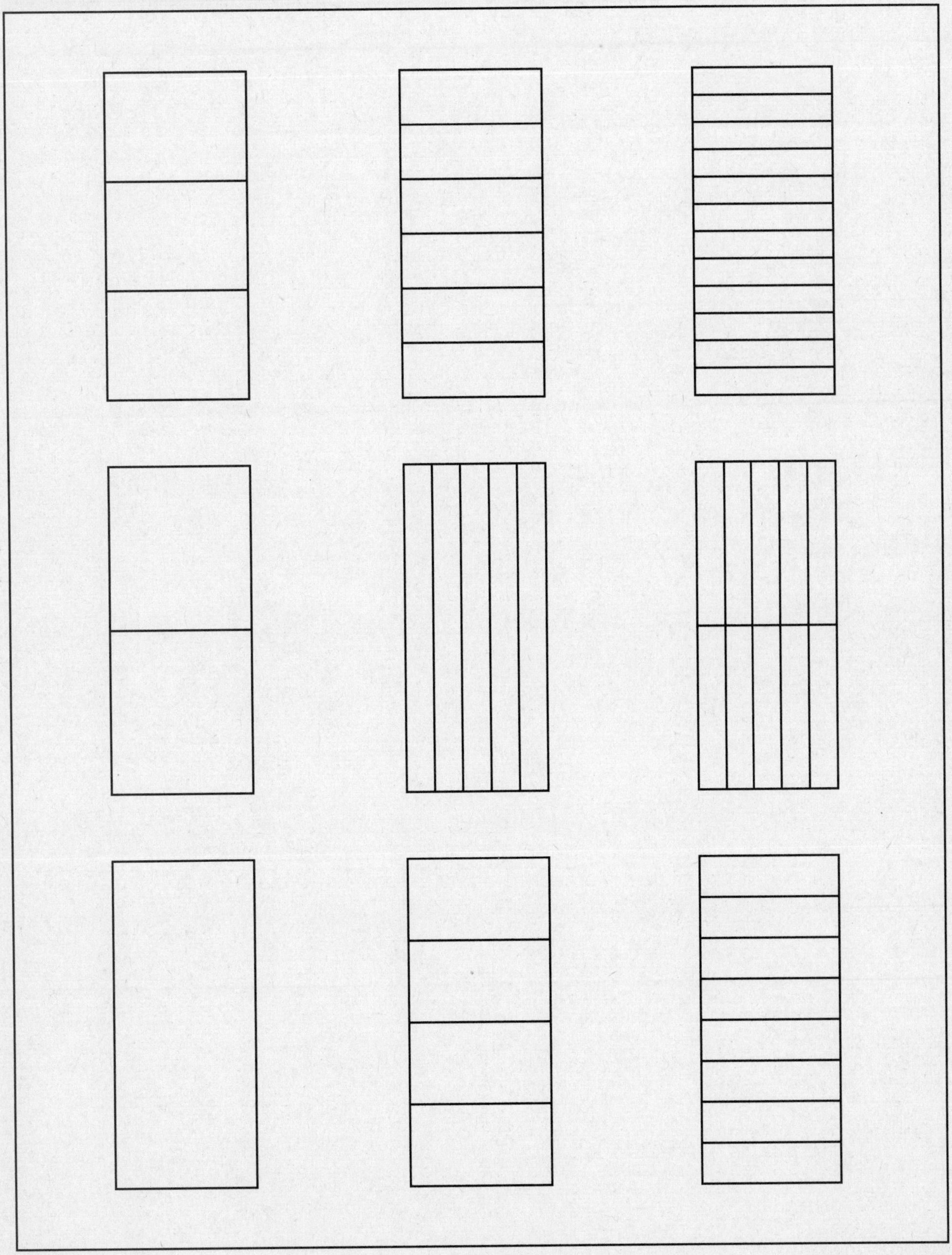